Silicon Nanowire Transistors

Ahmet Bindal • Sotoudeh Hamedi-Hagh

Silicon Nanowire Transistors

Springer

Ahmet Bindal
Computer Engineering Department
San Jose State University
San Jose, CA, USA

Sotoudeh Hamedi-Hagh
Electrical Engineering Department
San Jose State University
San Jose, CA, USA

ISBN 978-3-319-27175-0 ISBN 978-3-319-27177-4 (eBook)
DOI 10.1007/978-3-319-27177-4

Library of Congress Control Number: 2015959712

Springer Cham Heidelberg New York Dordrecht London

Printed on acid-free paper

Springer International Publishing AG Switzerland is part of Springer Science+Business Media (www.springer.com)

This book is dedicated to my all-time mentor and friend Dr. Seiki Ogura whose constant encouragement to my well-being will never be forgotten.

Dr. Ahmet Bindal

I dedicate this book to my loving family who have always been a constant source of support and happiness.

Dr. Sotoudeh Hamedi-Hagh

Preface

When we started exploring the possibility of using Silicon Nanowire Transistors (SNT) for the next-generation VLSI technology, we were initially quite uncertain in its outcome. The early device simulations did not reveal superior device performance compared to FinFETs. Besides, there was an issue of channel doping. Even high concentrations of Arsenic (and Boron) were able to replace only several dopant atoms in the MOSFET body. This created a serious problem in three-dimensional device simulations. The sequence of failures and disappointing results motivated us to look at alternative device designs. We used intrinsic silicon for the body of the device and changed the gate material from conventional polysilicon to metal to be able to adjust the gate work function and the threshold voltage. This approach also helped to eliminate short channel effects of the transistor; however, it also made the device fabrication steps in simulations more difficult due to the metal gate. Previous annealing steps used after the gate deposition step could not be used once the metal gate was deposited. The heat management became a critical issue and required several changes in the fabrication in order to form the vertical gate structure.

It was not until we created the level 6 SPICE models for n- and p-channel SNTs and used them in basic digital gates, we could observe the real potential of SNTs in circuit performance and power consumption. Motivated primarily by the power consumption results, we subsequently replaced the level 6 models with the more accurate BSIMSOI models in the next phase of our research in designing analog and digital circuits.

The organization of chapters pretty much follows the progression of our five-year long research. Chapters 1 and 2 examine the device design and characteristics of SNTs with dual and single work-function metal gates. Each of these chapters studies and measures the circuit performance and power consumption of basic digital gates with extrinsic device parasitics. The layout area of each gate is also included in each chapter and compared with various digital gates built with FinFETs. Chapter 3 examines the BSIMSOI SPICE model and all the intrinsic and extrinsic parasitic components of SNTs. High-speed analog applications are studied in Chapter 4 where SNTs are used in a single-stage amplifier, a differential pair, and a multi-stage operational amplifier. Chapter 5 examines the Radio Frequency (RF) applications. In this chapter, we presented the front end of an RF receiver and a Voltage-Controlled Amplifier (VGA). In Chapters 6 through 9, SNTs are used in various mega cells and complex digital systems. A complete Static Random Access Memory (SRAM) design and its layout are studied in Chapter 6. A Field-Programmable-Gate-Array (FPGA) architecture, circuit characteristics and layout in Chapter 7, an Integrate-and-Fire Spiking (IFS) neuron in Chapter 8, and a complete Direct Sequence Spread Spectrum (DSSS) baseband transmitter design in Chapter 9 are given to fully understand the implications of using SNTs in large-scale digital systems.

We firmly believe that SNTs are good candidates for the future of VLSI once the inherent complexities of device fabrication are overcome.

Dr. Ahmet Bindal
Dr. Sotoudeh Hamedi-Hagh

Contents

About the Authors

Ahmet Bindal Ahmet Bindal received his M.S. and Ph.D. degrees in Electrical Engineering from the University of California, Los Angeles, CA. His doctoral research was on the material characterization and analysis of HEMT GaAs transistors. During his graduate studies, he was a research associate and technical consultant for Hughes Aircraft Co. In 1988, he joined the technical staff of IBM Research and Development Center in Fishkill, NY, where he worked as a device design and characterization engineer. He developed asymmetrical MOS transistors and ultra-thin Silicon-On-Insulator (SOI) technologies for IBM. In 1993, he transferred to IBM in Rochester, MN, as a senior circuit design engineer to work on the floating-point unit for AS-400 main frame processor. He continued his circuit design career at Intel Corporation in Santa Clara, CA, where he designed 16-bit packed multipliers and adders for the MMX unit for Pentium II processors. In 1996, he joined Philips Semiconductors in Sunnyvale, CA, where he was involved in the designs of instruction and data caches, and various SRAM modules for the Trimedia processor. His involvement with VLSI architecture also started in Philips Semiconductors and led to the design of the Video-Out unit for the same processor. In 1998, he joined Cadence Design Systems as a VLSI architect and directed a team of engineers to design self-timed asynchronous processors. After approximately 20 years of industry work, he joined the Computer Engineering faculty at San Jose State University in 2002. His current research interests range from nano-scale electron devices to nano-scale architectures and robotics. Dr. Bindal has over 30 refereed scientific publications and 10 invention disclosures with IBM. He currently holds three U.S. patents with IBM and one with Intel Corporation.

Sotoudeh Hamedi-Hagh Dr. Hamedi-Hagh received his Ph.D. from the University of Toronto, Canada, in 2004. He joined the Electrical Engineering Department at San Jose State University (SJSU) in 2005. His areas of research and expertise include high frequency modeling of semiconductor device structures and design of Radio Frequency, Analog and Mixed-Signal integrated circuits for wireless and wireline communication systems. Dr. Hamedi-Hagh has developed the Radio Frequency Integrated Circuits laboratory and curriculum at both graduate and undergraduate levels with over $0.5 M research funding and through close collaborations with industries. He has received several California State University (CSU) professional development grants, CSU Research Funds, Research, Scholarship and Creative Activity (RSCA) grants, SJSU Planning Council Grants, College of Engineering professional development grants, and Junior Faculty Career Development Grants. He is a founding member of SJSU Smart Technology and Computing Center for Complex Systems (STCCS). In 2016, he was appointed as the Mixed-Signal endowed chair of the Electrical Engineering department. Dr. Hamedi-Hagh has over 30 refereed scientific journal and conference paper publications in prestigious national and international institutes and societies. He received the best paper award at the Micronet Symposium in Quebec, Canada, in 2001 and the IEEE International Symposium on Personal, Indoor and Mobile Radio Communications in Barcelona, Spain, in 2004. Dr. Hamedi-Hagh has advised several hundred projects on design of integrated circuits and systems. He holds seven US and world patents on wireless circuits, systems and cryptography. His latest patent introduces suspendance® and trajectance® laws as alternatives to Kirchhoff's laws for circuit analysis.

Chapter 1
Dual Work Function Silicon Nanowire MOS Transistors

1.1 Device Design

1.1.1 Introduction to Design Process

In the past, there were several attempts to develop alternative technologies, including molecular technologies [1, 2], that were aimed to replace the current VLSI technology. However, conventional silicon-based technologies prevailed as solid choices over the newcomers for fabricating low power nano devices and circuits without sacrificing high performance. As today's chips require larger die areas to accommodate complex System-On-Chip (SOC) designs, reducing overall power dissipation has been accepted as the major design objective, replacing the need for faster circuit performance. Recent modeling studies in undoped, double-gated SOI MOS transistors revealed that these transistors could produce an order of magnitude less leakage current compared to conventional bulk silicon MOS transistors for achieving ultra-low power consumption [3]. However, fabricating ultra-thin transistors sandwiched between two gates with adjustable work function is highly questionable in a production environment since both gates have to be made out of metal in order to produce proper threshold voltage and therefore to maintain a healthy circuit operation. Other studies on this device showed the effect of body thickness to alter the threshold voltage [4] and the variations of the back oxide thickness to decrease overall power dissipation [5]. Two-dimensional analytical modeling [6] and quantum mechanical modeling [7, 8] were also performed to better predict this device's performance and leakage current under different biasing conditions.

Another good candidate is a nano-scale, triple-gated SOI transistors or FINFETs. Theoretical studies conducted on these transistors explored the possibility of increasing transistor performance without increasing power consumption [9]. Recent experimental studies showed close-to-ideal subthreshold slope and Drain-Induced-Barrier-Lowering (DIBL) [10], both of which are important factors to reduce OFF current and power consumption [11]. Besides these promising

A. Bindal, S. Hamedi-Hagh, *Silicon Nanowire Transistors*,
DOI 10.1007/978-3-319-27177-4_1

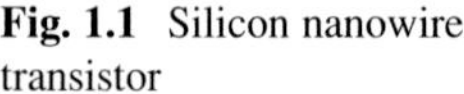
Fig. 1.1 Silicon nanowire transistor

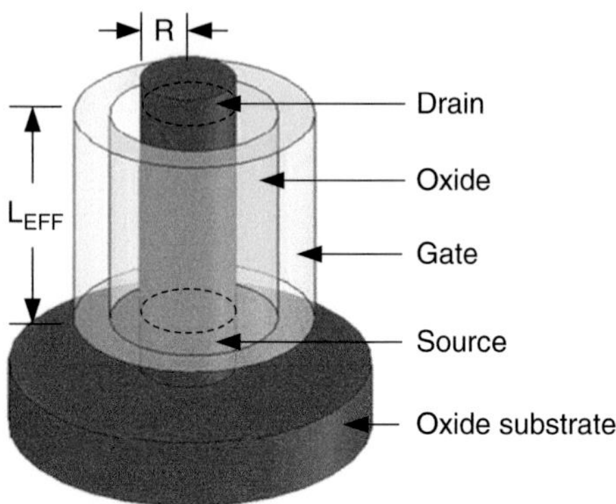

technologies, Silicon Nanowire MOS Transistors (SNT) also offers significant reduction in static and dynamic power consumption and compact layout area without sacrificing circuit performance. SNTs are vertically built on Silicon-On-Insulator (SOI) substrate and cylindrical in shape with a gate surrounding the entire perimeter of the transistor body as shown in Fig. 1.1. The source contact is placed at the bottom of the cylindrical body standing on the SOI substrate while the drain is placed at the top of the device interfacing the first metal layer. The primary objective of this chapter is to design SNTs with dual work function gates and use them in ultra-compact digital CMOS circuits that dissipate minimal static and dynamic power but perform equally or better than the state-of-the-art CMOS circuits.

The design criteria for each NMOS and PMOS SNT are outlined below:

1. NMOS and PMOS transistors need to have 300 mV threshold voltage for 1 V CMOS circuit operation for good noise immunity and low OFF current; therefore, different gate metals (dual work function) need to be used for each transistor.
2. The static OFF current has to be under 1 pA in each transistor.

However, to achieve these objectives requires changing the body geometry (channel length and radius) of both transistors and determining the device characteristics at each change.

This chapter also discusses the strengths and weaknesses of this technology. Low static and dynamic power dissipation, suppression of Short Channel Effects (SCE), and surface mobility enhancement may be considered as the advantages of SNTs. Alternative placement of NMOS and PMOS transistors as a crossbar configuration may also be counted as an advantage to simplify the layout; however, this method also increases the layout area. Other issues, such as source and drain contact resistances due to small body radius, source contact extension producing high source resistance, and fixed body dimensions resulting in non-adjustable ON currents (therefore limiting transient performance), are definite disadvantages of this technology.

One can trace the foundations of silicon nanowire technology in much earlier studies that investigate material properties and circuits. Silicon nanowires grown by Vapor–liquid–Solid (VLS) mechanism [12] and Chemical Vapor Deposition

(CVD) [13, 14] can be used to fabricate vertical SNTs. In fact, it was demonstrated that silicon nanowires could be used in Static Random Access Memory (SRAM) [15] and high-speed logic circuits [16]. Theoretical studies investigated the bulk and transport properties of silicon nanowires [17] and device properties as a function of wire diameter [18]. Circuit performance and power dissipation of SNTs were briefly studied in 3-D DNA architectures [19, 20].

1.1.2 The Criteria for Low Static Power Dissipation

There are three major components that result in low static power dissipation:

1. Junction leakage
2. Subthreshold leakage
3. Gate-Induced-Drain-Leakage (GIDL) current

Junction leakage current primarily depends on DIBL factor as shown in Eq. 1.1.

$$\mathrm{DIBL} = \left| \frac{V_{TSAT} - V_{TLIN}}{(V_{DS} = V_{DD}) - (V_{DS} = 50\mathrm{mV})} \right| \tag{1.1}$$

where V_{TLIN} and V_{TSAT} are the threshold voltages at $V_{DS} = 50$ mV and $V_{DS} = V_{DD}$, respectively.

Subthreshold leakage current is a function of subthreshold slope, S, and saturation threshold voltage, V_{TSAT}, as expressed in Eq. 1.2.

$$I_{SUB} = Io.10^{\frac{-V_{TSAT}}{S}} \tag{1.2}$$

Here, Io is the drain current at $V_{GS} = V_{TLIN}$ and S is given in Eq. 1.3 [21].

$$S = \frac{kT}{q} \log\left(1 + \frac{C_D}{C_{OX}}\right) \tag{1.3}$$

In this equation, C_D and C_{OX} are the channel depletion region and gate oxide capacitances, respectively.

The third component, GIDL current, is a strong function of transverse electric field, ES, at the semiconductor surface perpendicular to the device axis as given by Eq. 1.4 [22].

$$\mathrm{IGIDL} = A.E_S.\exp\left(-\frac{B}{E_S}\right) \tag{1.4}$$

where

$$E_S = \frac{V_{DG} - V_{FB} - 1.2}{3tox} \tag{1.5}$$

A is pre-exponential constant, B is a physically based exponential parameter suggested by [23], V_{DG} is the drain-to-gate potential, V_{FB} is the flat band voltage, and tox is the oxide thickness.

Therefore, the OFF current can be reduced by decreasing DIBL, tox, body doping concentration, and E_S. In this work, tox is set to minimum value of 1.5 nm to maintain the gate leakage current to a negligible level with respect to I_{OFF} as suggested by [3]; the body doping concentration is reduced to intrinsic level to minimize C_D, and E_S is also kept small due to non-overlapping gate-drain region and sub-10 nm wire radius [22].

1.1.3 Device Structure

Both NMOS and PMOS transistors are designed as enhancement type with uniform, undoped silicon bodies constructed perpendicular to the substrate. Both have the same body radius and effective channel length. Source/drain (S/D) contacts are assumed to have ohmic contacts. Both NMOS and PMOS transistors have metal gates and 1.5 nm thick gate oxide.

Device simulations are performed using Silvaco's 3-D ATLAS device simulation environment with a 1 V power supply voltage. Half of the device is constructed in a 2-D platform and then rotated around the y-axis to create a 3-D cylindrical structure for simulations. The device radius is changed from 1 nm to 25 nm while its effective channel length is varied between 5 nm and 250 nm.

1.1.4 Physical Models Used in Device Simulations

Even though sub-100 nm device geometry requires inclusion of Schrödinger's equation to calculate effective electron/hole masses and density of states due to the perturbations in the silicon conduction and valance bands, ATLAS simulator is limited to the full usage of such quantum mechanical effects. Instead, this study follows a semiclassical approach in which the semiconductor surface potential and density of states are corrected using density gradient method [24].

Mobility models are composed of two parts to estimate the effects of low and high electric fields. Lombardi's vertical and horizontal electric field dependent mobility model is used for low electric field effects [25]. Velocity saturation and high electric field effects are estimated by Caughey's drift velocity model [26]. Mobility degradation due to lattice temperature is included using Arora's model [27].

Concentration-dependent Shockley–Read–Hall recombination and surface recombination models are included to estimate the recombination rates in the bulk and at the silicon/oxide interface, respectively. Serberherr's impact ionization model constitutes the only generation model in the simulations [28]. Gate oxide tunneling mechanisms and hot carrier injection are ignored because these mechanisms largely depend on oxide growth and composition, and change from one processing condition to another.

1.1.5 Determining Metal Gate Work Function Values for NMOS and PMOS Transistors

The first task in this design process is to determine an individual metal work function for each NMOS and PMOS transistor at a minimum channel length of 5 nm in order to produce a threshold voltage of approximately 300 mV. This value constitutes 30 % of the 1 V power supply voltage and provides sufficient noise immunity for any CMOS gate. Threshold voltage of each NMOS and PMOS transistor is measured as a function of work function for the body radius from 1 nm to 25 nm as shown in Fig. 1.2. Longer channel length devices yield marginally higher threshold voltages and improve noise margin slightly. The intersection of threshold voltage with 300 mV level in Fig. 1.2 is projected to the x-axis to yield an individual work function value for each NMOS and PMOS transistor at a different body radius. Threshold voltages are measured using two different methods: the first method extrapolates the maximum slope of I_D–V_{GS} curve towards V_{GS}-axis and

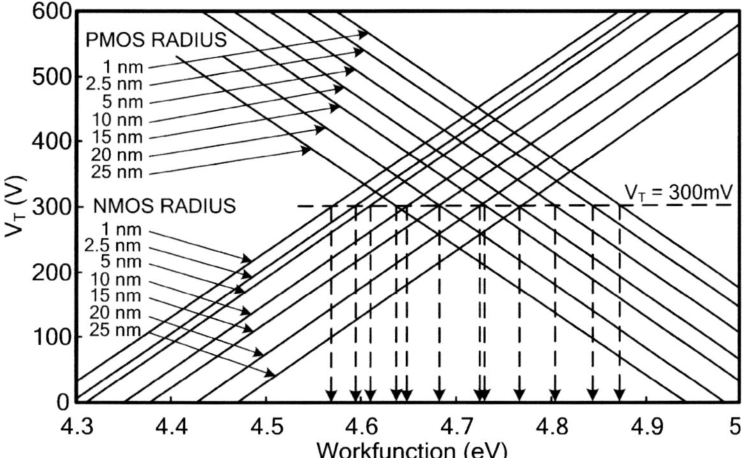

Fig. 1.2 Threshold voltages of NMOS and PMOS nanowire transistors as a function of metal work function at a minimum effective channel length of 5 nm. Radius of both NMOS and PMOS transistors is changed between 1 and 25 nm

defines the intercept as the threshold voltage; the second method determines the threshold voltage from the gate voltage at $I_{DS} = \zeta(W/L)$ for $V_{DS} = 50$ mV, where ζ is 10^{-7}A for NMOS and 10^{-8}A for PMOS transistors. This method is suggested by Liu et al. [29] and consistently produced 11 % and 3 % lower threshold voltages for NMOS and PMOS transistors, respectively.

1.1.6 The OFF Current Requirement

The leakage current is an important factor towards lowering overall standby power consumption; both NMOS and PMOS transistors are designed to have static leakage currents smaller than 1 pA, which is significantly smaller than SOI transistors in earlier modeling studies [3, 5, 9] and several orders of magnitude smaller than the technology trend predicted by Sery et al. [30]. Therefore, while most transistors with 1 nm to 5 nm radius produced I_{OFF} less than 1 pA and were considered as potential candidates for an optimum transistor design, transistors with larger radii were eliminated because their leakage currents exceeded 1 pA as shown in Fig. 1.3. In this figure, the transistor geometries closest to the dashed line are considered potential candidates since they produce higher ON currents for a given value of I_{OFF}.

1.1.7 Intrinsic Transient Time

Following the device selection process for low power dissipation in Fig. 1.3, the intrinsic transient time, τ, of each "selected" transistor is measured and then plotted as a function of I_{ON} in Fig. 1.4. Intrinsic transient time determines the time interval for a transistor to charge (or discharge) the gate capacitance of an identical transistor when it is fully on ($V_{DS} = V_{GS} = 1$ V) and it is a quick way of understanding the transient characteristics of an individual transistor without building any circuitry. In Fig. 1.4, ON currents of the selected NMOS and PMOS transistors start diverging from each other after 4 nm radius and 40 nm effective channel length; larger wire radius provides higher I_{ON} values for NMOS transistors, but it reaches a saturation plateau for PMOS transistors. Therefore, the 4 nm radius and 40 nm effective channel length combination is considered an optimal choice to produce approximately equal drive currents and intrinsic transient times for both NMOS and PMOS transistors.

Figure 1.5 shows the ON currents for NMOS and PMOS transistors as a function of L_{EFF} for different wire radii and helps to explain the ON current behavior of each device in Fig. 1.4. For wire radius greater than 5 nm, ON currents of both NMOS and PMOS transistors increase with decreasing L_{EFF} as shown in Eq. 1.6 [31, 32].

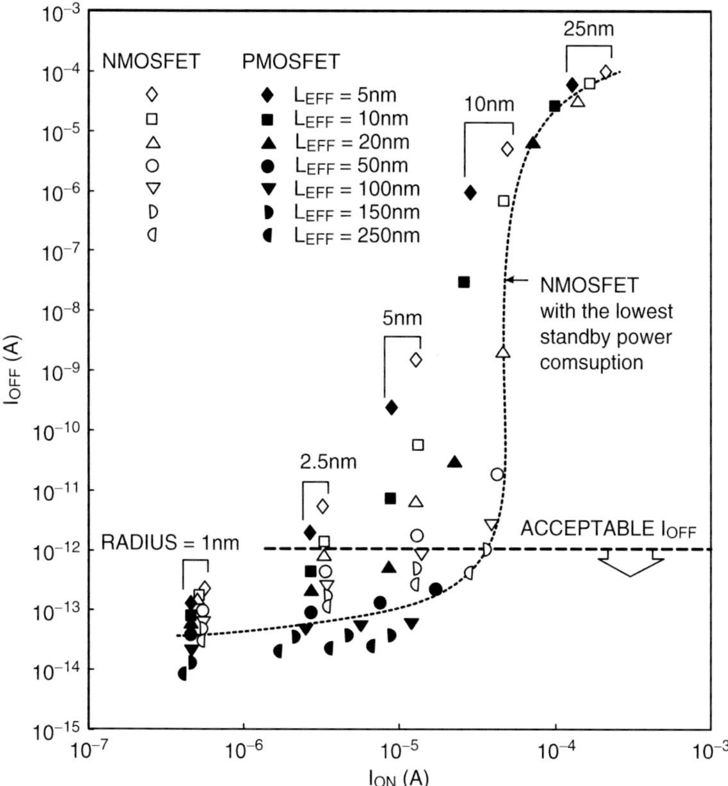

Fig. 1.3 The ON versus OFF current of NMOS and PMOS nanowire transistors. The radius of both transistors is changed between 1 and 25 nm while their effective lengths are varied between 5 and 250 nm. Each transistor has a specific gate work function value for each radius as specified in Fig. 1.2. Note that transistors with shorter effective channel lengths produce higher OFF currents

$$I_{ON} = \frac{\mu_{EFF}.\varepsilon ox.W}{2L_{EFF}.t'ox}\left\{(V_{GS}-V_T)^2 - \frac{16k^2T^2}{q^2}.\frac{t'ox.R.\varepsilon s}{\varepsilon ox}.\exp\left[\frac{q}{kT}(V_{GS}-V_{TO}-V_{DS})\right]\right\} \quad (1.6)$$

where

$$t'ox = R.\ln\left(1+\frac{tox}{R}\right) \quad (1.7)$$

$$V_{TO} = V_{FB} + \frac{kT}{q}\ln\left(\frac{kT\varepsilon sN}{q^2 ni^2}\right) \quad (1.8)$$

Here, N and R are the body doping concentration and radius of the SNT, respectively.

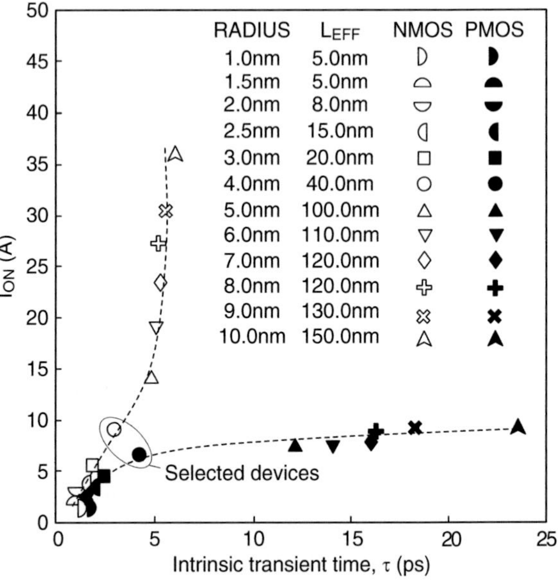

Fig. 1.4 The ON current versus intrinsic transient time of the "selected" NMOS and PMOS nanowire transistors whose leakage currents are below 1 pA

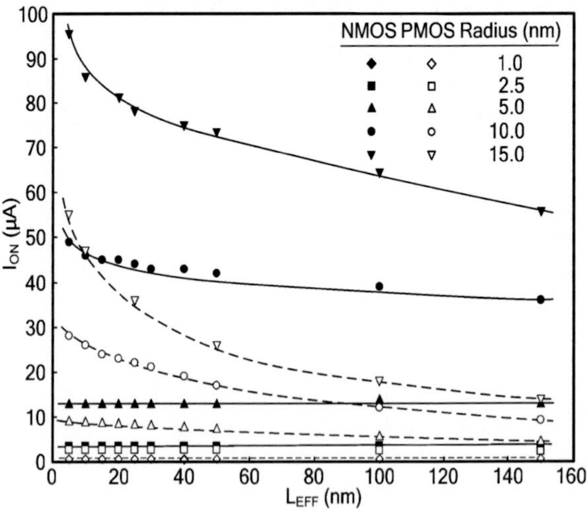

Fig. 1.5 The ON current of NMOS and PMOS silicon nanowire transistors as a function of effective channel length and body radius. Each transistor has a specific gate work function value for each radius as specified in Fig. 1.2

However, as the wire radius is further reduced from 5 nm to 1 nm, ON currents become independent of L_{EFF}. This behavior is not supported by the expression in Eq. 1.6. If inversion charge concentration, Qi, and drift velocity of electrons (holes) are examined throughout the body of small radius devices, one observes that both charge distribution and velocity are uniform. For example, if $V_{GS} = V_{DS} = 1$ V is applied to an NMOS transistor whose radius is smaller than 5 nm, the value of Qi approaches 10^{19} cm^{-3} and electron drift velocity becomes equal to 10^7 cm/s for device lengths between 5 and 150 nm. These observations suggest that electrons travel with saturation drift velocity across the transistor body and becomes independent of L_{EFF} as given by Eq. 1.9 [21].

$$I_{ON} = \pi.R^2 vsatQi \tag{1.9}$$

The validity of this statement can be further verified by computing the ON current ratios of small radius devices in Fig. 1.5 and comparing them against the square of body radius ratios. For example, I_{ON} of 1 nm, 2.5 nm, and 5 nm radius NMOS transistors are 0.54 μA, 3.3 μA, and 13 μA, respectively. When we compute the ratio of the ON currents of the R = 5 nm device to the R = 2.5 nm device, we obtain 3.94. If the same ratio is computed using Eq. 1.9, the result becomes equal to 4, assuming Qi in both devices is equal. Similarly, the ratio of ON currents of the R = 2.5 nm device to the R = 1 nm device produces 6.11 from Fig. 1.5 while the same ratio produces 6.25 according to Eq. 1.9.

For large wire radius shown in Fig. 1.5, I_{ON} follows Eq. 1.6 and the ratio of ON currents becomes equal to the ratio of effective electron and hole mobilities in NMOS and PMOS transistors if minor deviations in threshold voltages are ignored. For example, default effective electron and hole mobilities in Silvaco's ATLAS design simulation environment produce an ON current ratio of 3.9, whereas the ON current ratio extracted from Fig. 1.5 is equal to 3.38 for 10 nm radius NMOS and PMOS transistors. However, as the wire radius is reduced and the transistor bulk effect disappears, the ON current follows Eq. 1.9 and the ratio of ON currents becomes approximately proportional to the square of the NMOS and PMOS transistor wire radius. For example, 1 nm wire radius NMOS and PMOS transistors produce 0.54 μA and 0.48 μA ON currents, respectively. The ratio of ON currents approaches unity rather than approaching to the ratio of effective electron to hole mobilities as in large radius devices.

1.1.8 DC Device Characteristics

Figure 1.6 shows the threshold voltage roll-off of the 4 nm radius transistors with effective channel lengths ranging between 40 nm and 150 nm. The figure also includes earlier bulk and SOI transistor data for comparison [33–37]. The amount of ΔV_T is 6 mV for the NMOS and 11 mV for the PMOS transistors. These results are more than an order of magnitude smaller than the values of bulk silicon transistors which require heavily doped substrates to prevent SCE but consequently suffer from early impact ionization and large leakage currents.

The threshold voltage behavior of SNTs in Fig. 1.6 is not surprising because the same trend can also be seen in Fig. 1.7. This figure illustrates that the SCE gradually disappears as wire radius decreases towards 1 nm; larger radius devices are affected by the SCE and exhibit in excess of 100 mV threshold voltage change. This shows that bulk transistors or transistors fabricated on an SOI substrate thicker than 5 nm are still susceptible to threshold voltage variations as a function of device geometry.

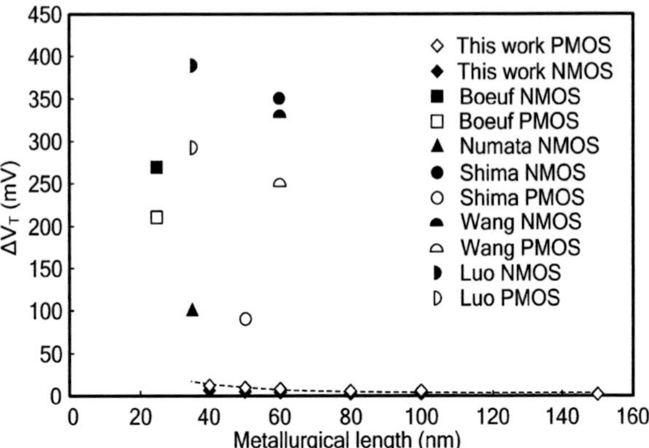

Fig. 1.6 Threshold voltage roll-off characteristics of undoped, dual work function NMOS and PMOS nanowire transistors at a 4 nm body radius. Prior work is included for comparison

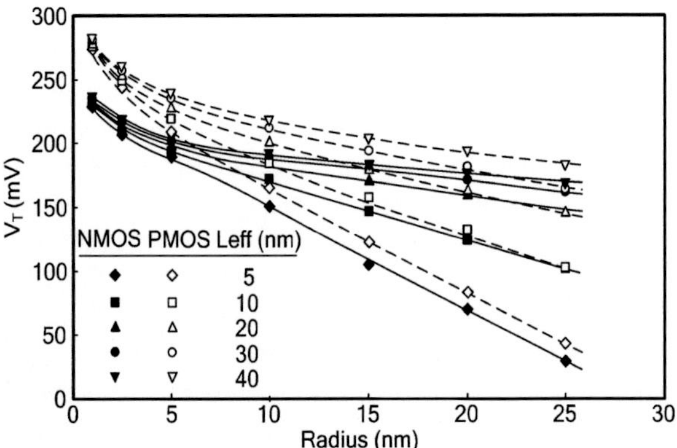

Fig. 1.7 Threshold voltage of undoped, dual work function NMOS and PMOS nanowire transistors as a function of radius and effective channel length. All NMOS transistors have 4.5 eV and all PMOS transistors have 4.9 eV metal gate work function. Short channel effects decrease as channel length is reduced

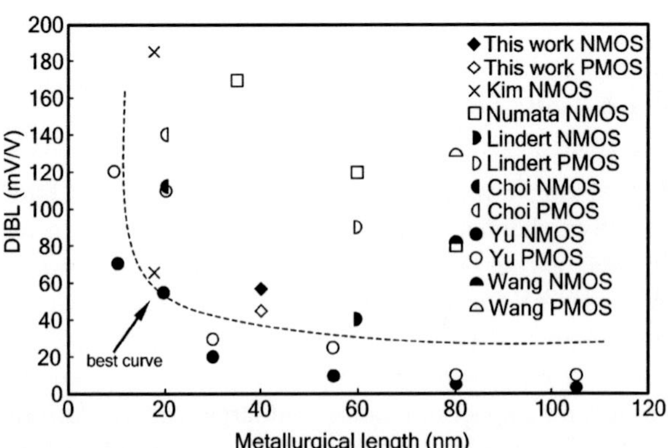

Fig. 1.8 DIBL of undoped, dual work function NMOS and PMOS nanowire transistors with 4 nm body radius and 40 nm effective channel length. Prior work is included for comparison

Silicon wire transistors, having a radial gate configuration, controls and suppresses SCE simply by reducing wire radius.

The amount of DIBL is 57 mV/V for the NMOS and 53 mV/V for the PMOS transistors with 4 nm radius and 40 nm effective channel length. These values are shown in Fig. 1.8 and compared with previously published data [3, 10, 34, 36, 38, 39].

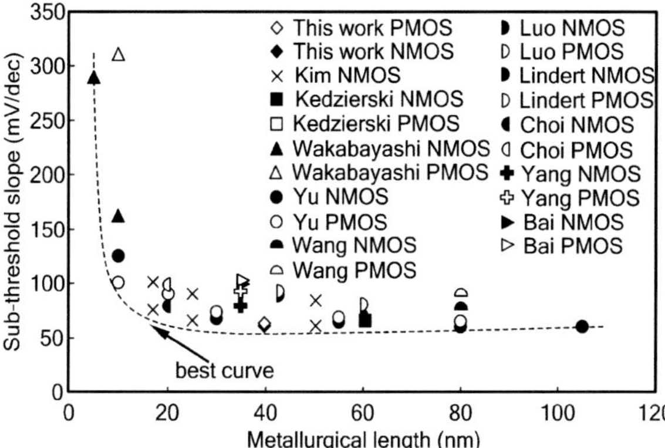

Fig. 1.9 Subthreshold slope of undoped, dual work function NMOS and PMOS nanowire transistors with 4 nm body radius and 40 nm effective channel length. Prior work is included for comparison

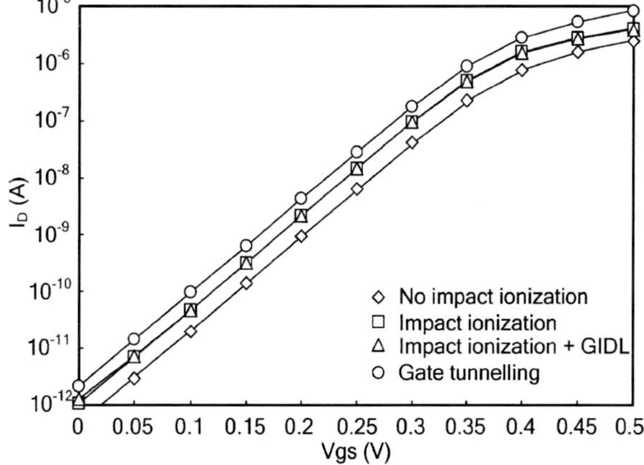

Fig. 1.10 The OFF current components for 4 nm radius and 40 nm effective channel length NMOS transistor

Subthreshold slope is 62 mV/dec for NMOS and 62.5 mV/dec for PMOS transistors at a drain voltage of 1 V. These results are plotted in Fig. 1.9 and show close-to-ideal characteristics in comparison with Kim's modeling results on double-gated SOI transistors [3] and previously published experimental data [10, 36–43].

Figure 1.10 shows the three components of the OFF current for NMOS transistor. Subthreshold leakage is the first component when there is no impact ionization in the device body. Junction leakage is the result of impact ionization at high lateral electric fields and it doubles the total OFF current. Band-to-band leakage or GIDL component is small compared to junction leakage because of three factors. The first factor is the absence of a gate-drain overlap region in the proposed device structure: only fringing component of the transverse electric field emanating from the edge of the gate may induce GIDL. The second factor is the decrease in transverse electric field with respect to a bulk device with a single gate: surface potential in a bulk or partially depleted SOI device is appreciable to promote GIDL current generation [22]. The third factor is the magnitude of the power supply voltage: the drain-to-gate potential being less than the silicon band gap is not an effective way to create enough band-bending at the semiconductor surface to allow the valance band electrons to tunnel into the conduction band. Gate oxide tunneling using Concannon's model [44], on the other hand, may increase the OFF current beyond its designed limit as it almost doubles the total OFF current as shown in Fig. 1.10. However, the magnitude of this current primarily depends on processing conditions including gate oxide composition, quality and defect levels during growth, and it is not included in this study.

Figure 1.11 shows the ON current of an NMOS transistor with and without quantum approximations. Van Dort's model is designed for thin gate oxide devices and empirically corrects the surface potential by broadening the energy band gap [45]. Density gradient model is calibrated with the Poisson–Schrodinger equation in simulations and it calculates position-dependent potential energy from the semiconductor surface towards the current transport axis according to higher derivatives of carrier distribution in the channel [24]. Potential energy corrections consequently modify electron and hole distributions in the channel and compute electron and hole current densities.

Figure 1.12 shows the output I–V characteristics of NMOS SNTs whose radius values are between 1 nm and 25 nm at $L_{EFF} = 40$ nm. The drain–source breakdown

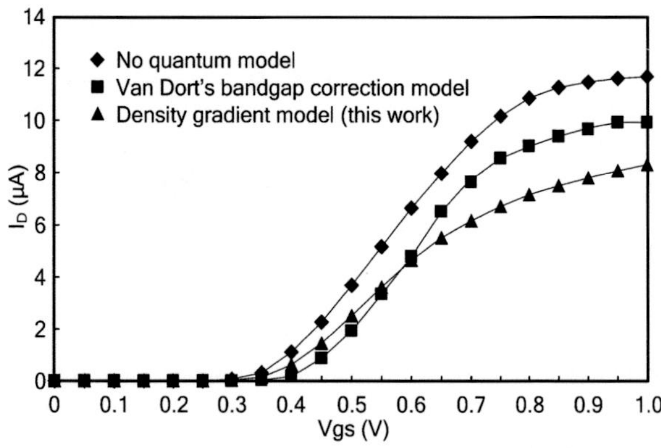

Fig. 1.11 ON currents with and without quantum models for 4 nm radius and 40 nm effective channel length NMOS nanowire transistor

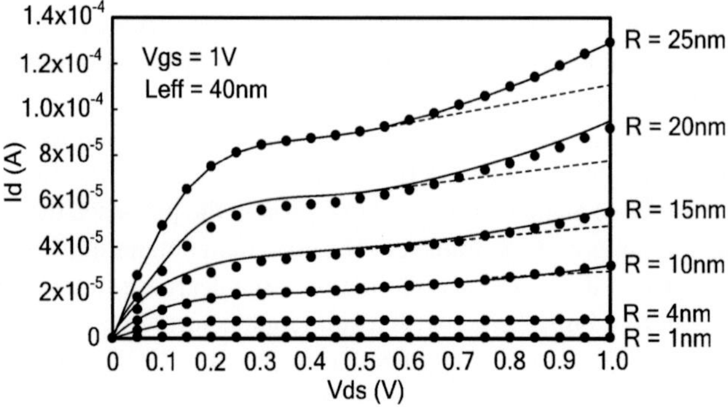

Fig. 1.12 Output I–V characteristics of NMOS nanowire transistors with an effective channel length of 40 nm at $V_{GS} = 1$ V. Each nanowire transistor has a leakage current below 1 pA. Dashed lines show the output I–V characteristics of nanowire transistors with "virtual" body contacts

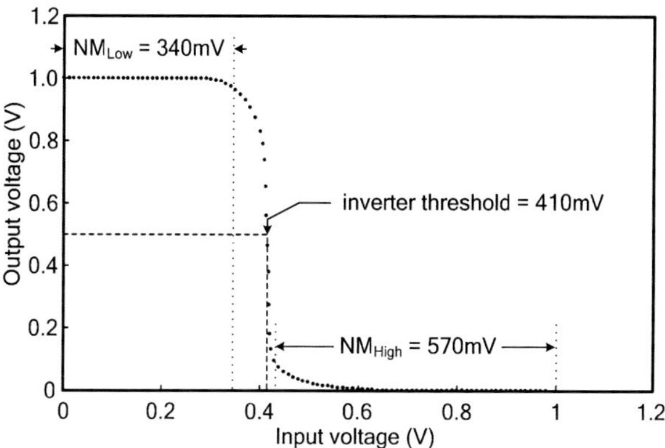

Fig. 1.13 Transfer curve of an inverter composed of 4 nm radius and 40 nm effective channel length undoped, dual work function NMOS and PMOS nanowire transistors. The projection of 0.5 V output voltage onto x-axis indicates 410 mV inverter threshold voltage

voltage moves towards higher values and finally disappears as radius is reduced towards 1 nm. Note that the breakdown voltage is a function of ionized electron–hole pairs throughout the device body, and it forms a "soft" kink effect in each I–V curve. The kink effect disappears when the transistor bulk is connected to a "virtual" ground in simulations as indicated by the dashed lines in Fig. 1.12.

Surface mobility enhancement in silicon nanowire transistors is also investigated. Semiconductor surface potential in a radial transistor body diminishes as the body radius is reduced. Surface potential in bulk transistors, on the other hand, reaches its maximum value because of the substrate contact and degrades the device mobility. An NMOS transistor with 4 nm radius and 40 nm effective channel length produces an ON current of 8.3 μA, whereas a bulk NMOS transistor with the same body configuration produces only 6 μA.

Figure 1.13 illustrates the first circuit-related result and shows the inverter transfer characteristics produced by 4 nm radius and 40 nm effective channel length NMOS and PMOS SNTs. The inverter threshold voltage computed by the projection of $V_{OUT} = 0.5$ V to the x-axis is slightly off-center at 410 mV due to the slightly higher NMOS drive current, but the inverter still produces sufficient low and high noise margins at 340 and 570 mV, respectively, for noise-free circuit operation.

1.2 Circuit Simulations and Performance

1.2.1 Parasitic Extraction and Post-layout Issues

To understand the circuit performance, power dissipation, and layout, several primitive gates, including an inverter, 2-input and 3-input NAND, NOR, XOR gates, and a full adder were built. All measurements were conducted before and after parasitic layout extraction and compared with each other to understand the effects of parasitic wire resistance, capacitance, and contact resistance on circuit performance. Since these transistors are constructed perpendicular to the substrate,

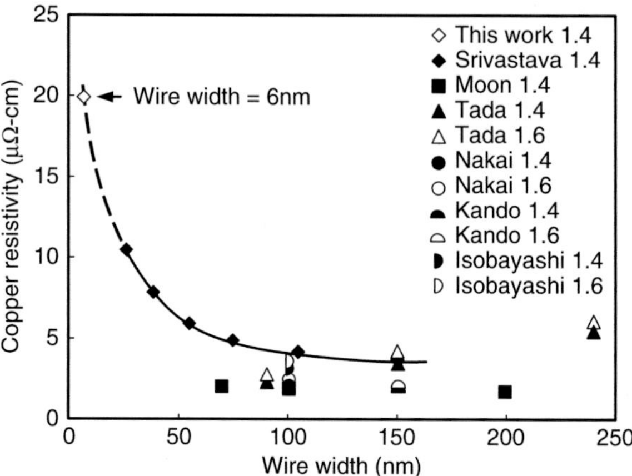

Fig. 1.14 Copper resistivity as a function of width. Srivastava's scattering model is extrapolated to obtain the resistivity value for 6.4 nm copper wires in this study. Prior experimental data are included for comparison

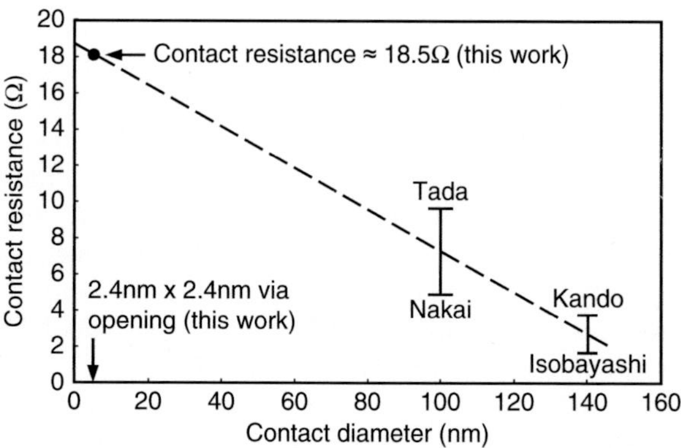

Fig. 1.15 Copper contact resistance as a function of contact diameter. The extrapolated value from earlier experimental data provides contact resistance value for this study

the minimum exposed transistor feature on the layout is 4 nm wire radius to make contacts. Copper wires with 6.4 nm width and 1.4 aspect ratio (wire height to width) are used for interconnects and 2.4 nm by 2.4 nm vias are used for contacts. Since sub-10 nm range copper wire electrical characteristics do not exist in the literature, copper resistivity was extrapolated from Srivastava's model on 1.4 aspect ratio wires [46]. Figure 1.14 shows copper resistivity as a function of wire width for aspect ratios of 1.4 and 1.6, and also contains experimental results for comparison purposes [47–51]; 20 μohm-cm resistivity was subsequently used to calculate the sheet resistance for 6.4 nm wide interconnects. Similarly, contact resistance was extrapolated from the experimental data for 100 nm and larger via diameters and resulted in 18.5 Ω for each metal contact as shown in Fig. 1.15 [48–51]. The estimations on contact resistance and wire resistivity likely contain errors since these parameters are extracted either from a technology that supports 100 nm wire features or extrapolated from a simplified scattering model that does not take into account crucial scattering mechanisms such as interface (wire surface) and grain boundary scattering [52]. Especially, 6.4 nm wire width is exposed to all such mechanisms because this dimension is below a typical grain size of copper (approximately 10 nm) and much lower than the mean free path of electrons (40 nm). N-well/P-well extension to form a source contact also introduces a series resistance to the transistor. The measured value of source extension for the NMOS transistor is 650 Ω and for the PMOS is approximately 2.5 kΩ. However, the change in overall circuit delay as a result of interconnect sheet resistivity, contact resistance and source extension is not significant because the equivalent SNT channel resistance is much larger than the sum of all these extrinsic resistances. A simple RC calculation on inverter rise and fall times reveals approximately 34 kΩ for PMOS and 7 kΩ for NMOS transistor channel resistance. If one limits the total contact, wire and source extension resistances to be 10 % of the equivalent NMOS channel resistance (or 790 Ω) to avoid interconnect-related delays, then the discharge path can accommodate 90 nm long copper wire between two contacts. The maximum wire length in the inverter layout is less than 50 nm long. The inverter charge path can even support more wire resistance since the equivalent PMOS channel resistance is 34 kΩ instead of 7 kΩ. More complex circuits containing multiple transistors in series can tolerate higher number of contacts and wire lengths in the charge and discharge paths. However, wire lengths outside the cell boundary in the form of long chip-level routes have the greatest sensitivity to resistivity errors and become an important issue when considering overall circuit delays and slow logic transitions. Fortunately, in the upper-metal routing, design rules are more relaxed, allowing wider and thicker wires.

Area, fringe, and coupling capacitances of metal 1 and metal 2 wires per unit length are calculated using Ansoft's 2-D electrostatic solver. These capacitance values are used to extract metal-to-metal and metal-to-substrate parasitic capacitances from layouts for circuit simulations.

1.2.2 *Transient Performance*

Post-layout transient characteristics of various CMOS gates composed of 4 nm wire radius and 40 nm channel length transistors are shown in Figs. 1.16 and 1.17 in terms of worst-case transient time and worst-case delay, respectively. The worst-case transient time is determined by projecting 10 and 90 % of the output voltage onto the time-axis and measuring the difference. Similarly, the worst-case delay is determined by projecting 50 % of the input and 50 % of the output voltage values onto the time-axis and measuring the difference. Each transient characteristic is plotted as a function of the output capacitance. Considering the gate capacitance of a single transistor is 32 aF, maximum output capacitance of 200 aF in simulations corresponds to a fan-out of approximately six identical transistors.

The worst-case transient time is essentially equivalent to the rise time of a gate since a PMOS transistor has almost five times higher resistance compared to an NMOS transistor as discussed earlier. The worst-case transient times of the inverter, 2-input and 3-input NAND-gates in Fig. 1.16 overlap each other primarily due to

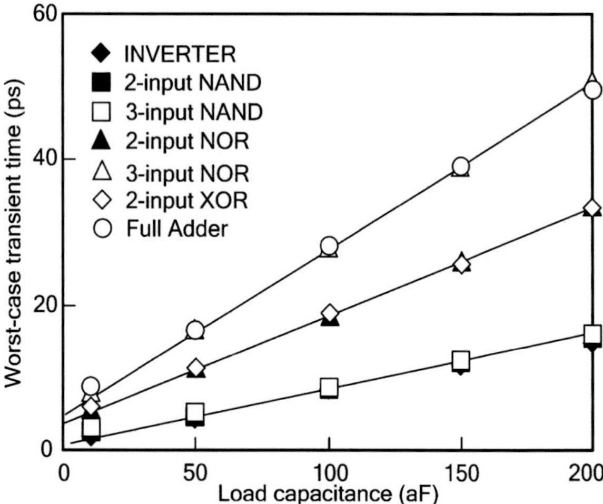

Fig. 1.16 Worst-case post-layout transient time characteristics of various primitive gates built with 40 nm effective channel length and 4 nm body radius NMOS and PMOS nanowire transistors

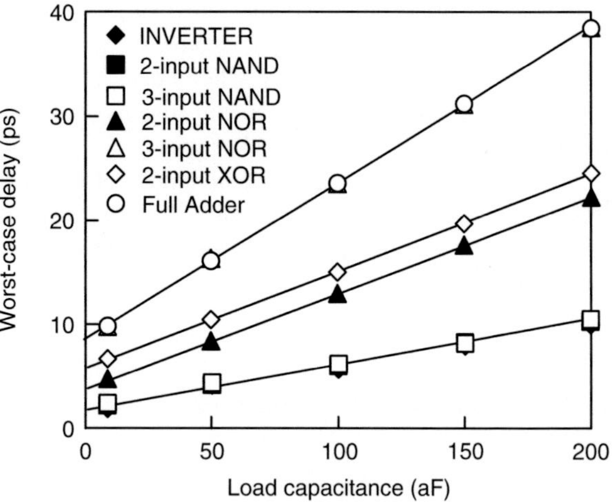

Fig. 1.17 Worst-case post-layout propagation delay characteristics of various primitive gates built with 40 nm effective channel length and 4 nm body radius NMOS and PMOS nanowire transistors

the single PMOS transistor charging the output capacitance. The worst-case transient times of the 2-input NOR and XOR circuits cluster together because there are two PMOS transistors in series charging the output load. The full adder and 3-input NOR circuits are in close proximity and reveal the highest transient times because the number of PMOS transistors in series increases from two to three in the critical charging path. The worst-case transient time of the full adder is expressed as $T = 4.74 + 0.23C_L$ in picoseconds, where C_L is the output load capacitance in aF.

The worst-case delay in Fig. 1.17 behaves similar to the worst-case transient time in Fig. 1.16 because the worst-case delay uses the same critical charging and discharging paths. The worst-case delay of the full adder circuit is expressed as $T_D = 8.50 + 0.15C_L$ in picoseconds.

The worst-case gate delay values of silicon nanowire technology are comparable to SOI but smaller than bulk silicon technologies. Kim et al. [3] obtained 4 ns and 5 ns individual inverter delays from a chain of double-gated SOI and bulk silicon inverters, respectively. The worst-case inverter delay in this study is approximately 2.5 ps when the inverter output is connected to the input of an identical inverter.

1.2.3 Power Dissipation

The worst-case power dissipation is composed of static power dissipation discussed earlier and dynamic power dissipation which is a function of frequency of operation, fop, power supply voltage, V_{DD}, and load capacitance, C_L as shown in Eq. 1.10.

$$\text{Pdyn} = \text{fop}.C_L.{V_{DD}}^2 \tag{1.10}$$

When V_{DD} and fop are adjusted to achieve the optimum circuit performance and noise margin, the only possible variable to reduce Pdyn is the load capacitance. Even though the dimensions of a bulk transistor can be changed to have the same gate capacitance of a single nanowire transistor, impact ionization, punch-through effect, and high S/D capacitance are still potential problems for the bulk device. Dual-gated SOI transistors benefit the same advantages as the SNTs, but their gate capacitance, and therefore the dynamic power dissipation, doubles as shown in Eq. 1.10.

The worst-case post-layout power dissipation of various logic gates with 10 aF capacitive load is shown in Fig. 1.18 as a function of frequency. The worst-case power dissipation is obtained by considering all possible input combinations to a logic gate, measuring the average value of the power supply current within one clock period for each combination (activity factor = 1 %) and finally selecting the combination that yields the maximum average current. Each current waveform is averaged in one clock period during charging and discharging the output capacitance. The worst-case power dissipation of a 2-input NAND gate is 36.9 nW at

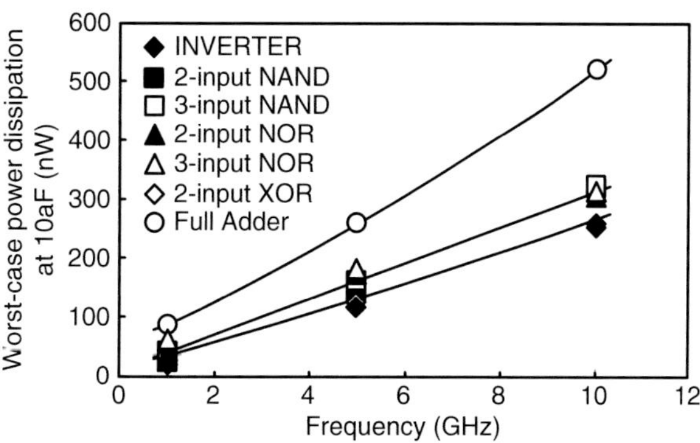

Fig. 1.18 Worst-case post-layout power dissipation of various primitive gates built with 40 nm effective channel length and 4 nm body radius NMOS and PMOS nanowire transistors at 10 aF capacitive load

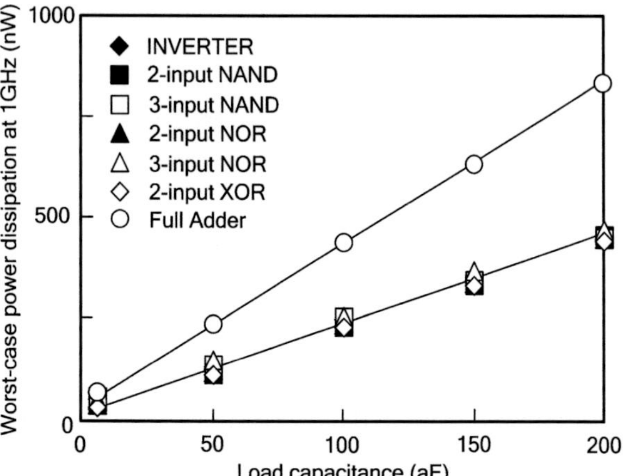

Fig. 1.19 Worst-case post-layout power dissipation of various primitive gates built with 40 nm effective channel length and 4 nm body radius NMOS and PMOS nanowire transistors at 1 GHz operating frequency

1 GHz and increases by 24.9 nW/GHz for a 10 aF output load. The worst-case power dissipation increases with number of transistors, layout complexity, and the number of "parallel" charging or discharging paths to a capacitive load. A full adder, as a more complex circuit, dissipates 85.0 nW at 1 GHz and the power dissipation increases by 51.2 nW/GHz for a 10 aF output load.

Figure 1.19 shows the worst-case post-layout power dissipation of each gate as a function of load capacitance at 1 GHz. The worst-case power dissipation for the 2-input NAND gate and full adder is $P = 14.1 + 2.19C_L$ and $23.6 + 4.04C_L$, respectively, in nanowatts.

1.2.4 Cell Layout and Gate Area Estimations

Layouts of various gates including an inverter, 2-input and 3-input NAND, NOR, and XOR circuits, and a full adder are designed using 4 nm radius and 40 nm effective channel length nanowire transistors. Figure 1.20 shows the cross section and the corresponding layout of an SNT. The active region defines the circular body of the SNT, which is surrounded by an N-well if the transistor is an NMOS device or P-well if it is a PMOS transistor. The outmost circle represents the metal gate. All contacts are indicated by 2.4 nm by 2.4 nm black squares touching the drain, the source and the gate of the transistor. Figure 1.21 shows the layout of a full adder. All interconnects between transistors are established by 6.4 nm wide wires. Further area reduction in this layout is possible if more than two metal layers are used to connect all its inputs and outputs with adjacent cells. Layout areas of the primitive gates used in this study are listed in Table 1.1. A recent 6-transistor SRAM cell designed in a 65 nm technology occupied a cell area of 0.57 μm^2 [41]. The 30-transistor full adder in this study has a cell area of approximately 0.11 μm^2, which is about 5 times smaller than the SRAM cell and contains five times more

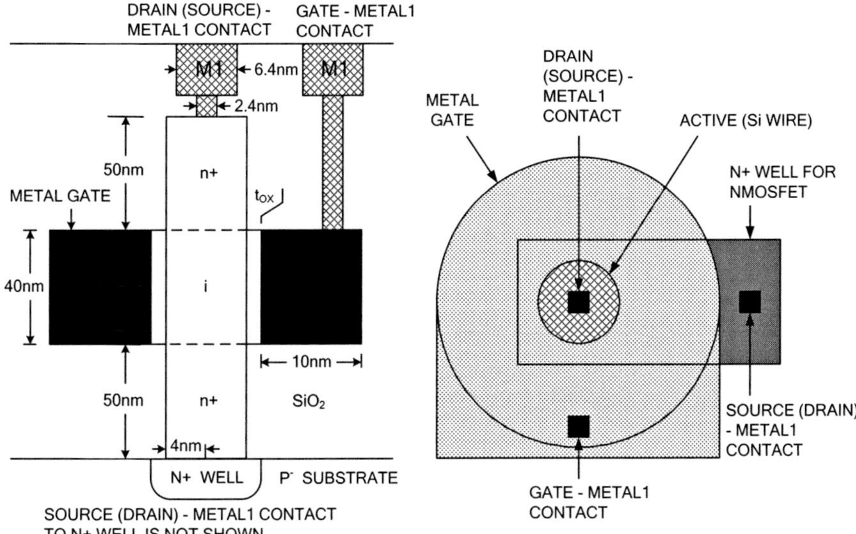

Fig. 1.20 Cross section and layout topology of a single, 40 nm effective channel length and 4 nm body radius NMOS transistor. Note that the source (drain) contact via is not shown on the cross section. A separate N-well completely surrounds the P-well of the PMOS transistor to prevent latch-up

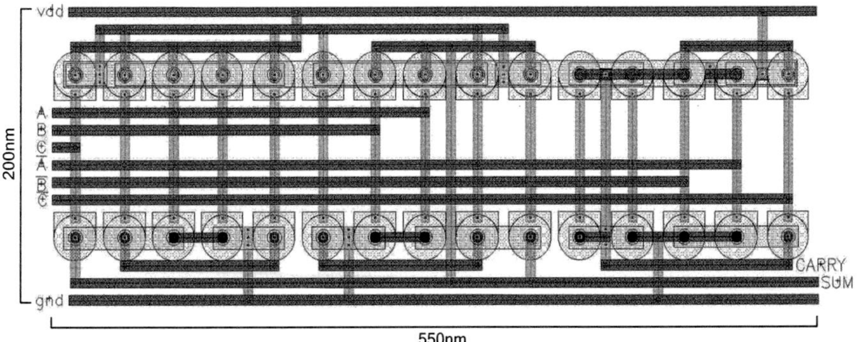

Fig. 1.21 The full adder layout using 40 nm effective channel length and 4 nm body radius NMOS and PMOS nanowire transistors. A, B, and C are the two inputs of the full adder and the carry-in, respectively. $\overline{A}$, $\overline{B}$, and $\overline{C}$ correspond to the two complemented inputs of the full adder and the complemented carry-in, respectively

transistors. There are two limiting factors which prevents further layout area reduction in SNTs: source contact extension and gate metal thickness. The former can be minimized by employing small contacts at the expense of increasing contact resistance; the latter has a thickness limit below which metallic grains separate from each other, forming a discontinuous metallic film. This study uses 10 nm gate metal thickness.

Table 1.1 Layout area of various gates built with 40 nm effective channel length and 4 nm body radius NMOS and PMOS nanowire transistors

Gate	Area (nm^2)
Inverter	7400
2-input NAND	14,800
3-input NAND	19,000
2-input NOR	14,800
3-input NOR	19,000
2-input XOR	24,000
Full adder	110,000

Table 1.2 Circuit performance, power dissipation, and layout area of full adder circuits in this study and earlier work

Lg (nm)	V_{dd} (V)	Fop (MHz)	P_T (nW)	Delay (ps)	Area (μm^2)	References
350	3.3	–	164,000	227	–	[55]
350	1.2	50	2490	2037	387	[53]
350	1.8	50	6090	827	387	[53]
350	2.5	50	12,820	528	387	[53]
350	3.3	50	24,120	406	387	[53]
350	3.3	–	65,000	400	–	[56]
250	3.3	–	58,000	300	–	[56]
180	3.3	–	30,000	100	–	[56]
180	1.0	100	2500	650	–	[57]
180	1.8	100	6230	292	100	[54]
180	1.0	100	1450	756	100	[54]
180	1.8	300	345	195	–	[58]
180	1.8	50	11	327	–	[59]
40	1.0	1000	556[a]	28[a]	0.11[a]	This work

[a]An output load of 130 aF (4 transistor gates)

A comparison of the SNT full adder and the earlier conventional adders is provided in Table 1.2 in terms of transient performance, power dissipation, and layout area [53–59].

1.2.5 *Manufacturability*

Silicon nanowire transistors can be manufactured relatively easier compared to the dual-gated SOI transistors. The processing steps in Fig. 1.22 shows a method to fabricate NMOS SNT using chemical–mechanical polishing and other conventional processing methods. Dual-gated SOI transistors require metal gates on both sides of the thin transistor body and their manufacturability may not be possible with conventional processing tools.

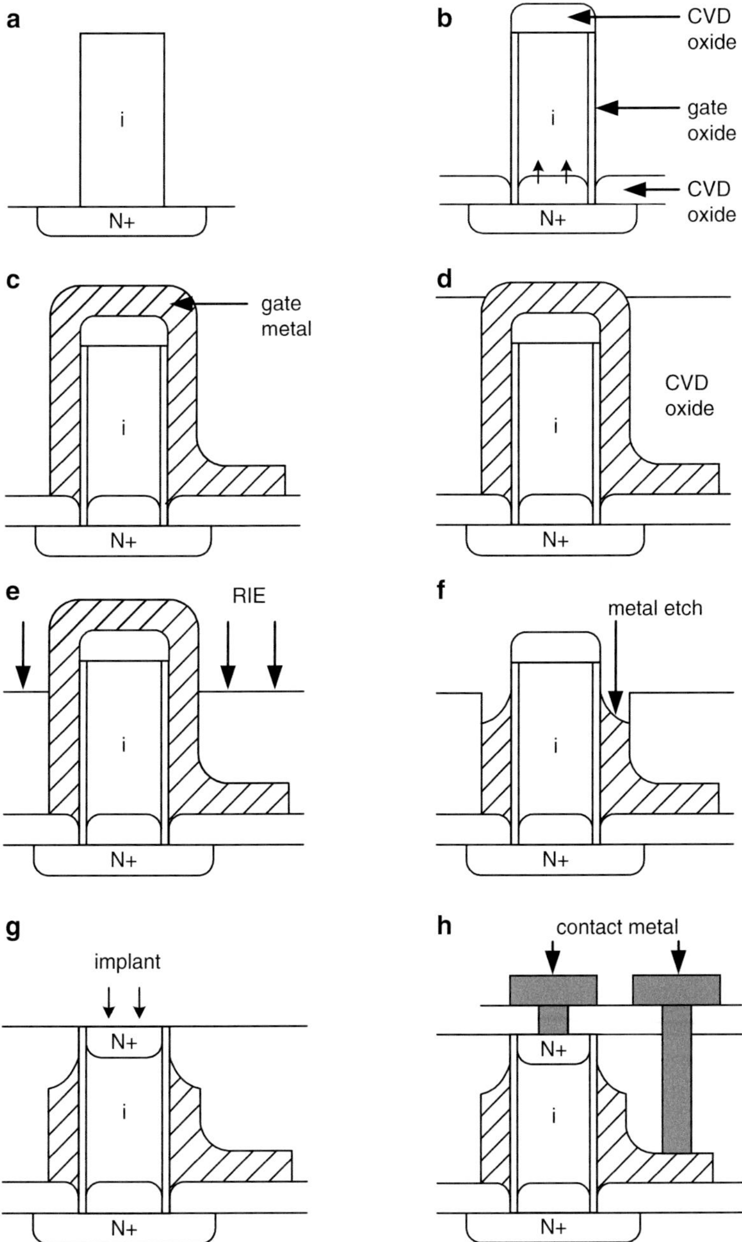

Fig. 1.22 Processing steps for NMOS transistor. (**a**) Intrinsic Si wire is grown, (**b**) Gate oxide is grown, and source junction is formed; anisotropic PECVD oxide is deposited, (**c**) Metal gate is deposited, (**d**) Thick oxide is deposited to the length of the wire and CMP is applied until wire end is detected, (**e**) Thick oxide is recessed to drain gate junction by preferential Reactive Ion Etching (RIE), (**f**) Metal gate is wet etched to define physical gate length, (**g**) Second thick oxide is deposited, CMP is applied, drain junction is formed by ion implantation, (**h**) Gate and drain contacts are formed

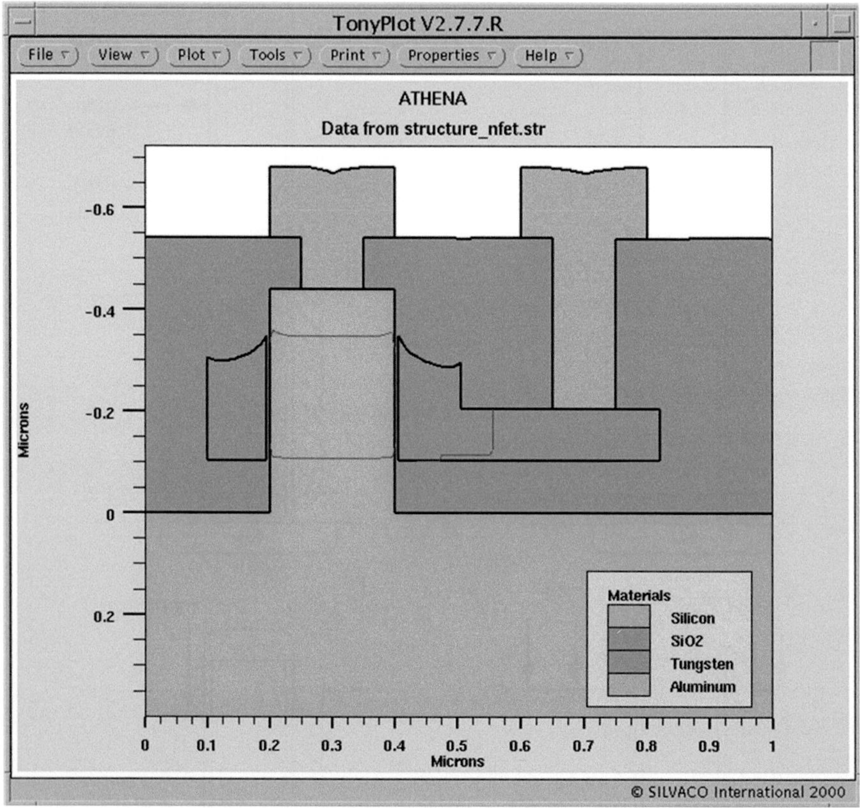

Fig. 1.23 Complete NMOS transistor cross section obtained by Athena process simulator

In order to verify the process flow in Fig. 1.22, two-dimensional processing simulations are carried out in Silvaco's Athena process design environment. The processing steps from Fig. 1.22a through Fig. 1.22h produced the NMOS transistor whose final cross section is shown in Fig. 1.23.

Initially, 450 nm long, intrinsic silicon nanowires are grown perpendicular to the heavily doped N-well (P-well) regions that define the source contacts for the NMOS (PMOS) transistor with Au catalyst. Both the phosphorus-doped N-well and boron-doped P-well define the source region of the NMOS and PMOS transistors, respectively.

After the wire growth, the Au catalyst is stripped from the top of the grown wire using a wet etch, as shown in Fig. 1.22a. This step is followed by a 5 nm gate oxide growth at 975 °C for 30 min. Next, 100 nm CVD oxide is deposited anisotropically to define the gate–source boundary [22]; anisotropic CVD oxide deposits only horizontally to the substrate, but it does not attach to silicon wire walls as shown in Fig. 1.22b. The oxidation step causes phosphorus atoms from the N-well (boron atoms from the P-well) to diffuse approximately 10 nm into the silicon wire from the substrate surface. A 100 nm thick tungsten layer is deposited as the gate material and defined as shown in Fig. 1.22c. Isotropic CVD oxide is deposited, and a Chemical–mechanical Polish (CMP) step is applied until tungsten is seen, as shown in Fig. 1.22d. CVD oxide is recessed to define the depth of the drain region or drain–gate boundary as shown in Fig. 1.22e. Exposed tungsten is wet etched, as shown in Fig. 1.22f. A second CVD oxide layer is deposited followed by a CMP step that stops at silicon. Low energy phosphorus (boron) is implanted perpendicular to the wire end to form the drain contact for the NMOS (PMOS) transistor as shown in Fig. 1.22g. A 20 s Rapid Thermal Annealing (RTA) step is applied at 900 °C to activate phosphorus (boron) implants without damaging the gate metal. A 100 nm CVD oxide is subsequently deposited. Aluminum via contacts are formed using a unidirectional Reactive Ion Etch (RIE) step. A 140 nm thick aluminum layer is deposited as the metal 1 layer as shown in Fig. 1.22h.

1.3 Summary

In this exploratory work, silicon nanowire CMOS circuits are studied for low power and high density VLSI applications. Three-dimensional undoped NMOS and PMOS nanowire transistors are designed and optimized in Silvaco's ATLAS device design environment to maximize the ON current and to keep the OFF current below 1 pA as a function of device geometry. Threshold voltage of each transistor is adjusted by an individual gate metal work function. As the body radius is reduced from 25 nm towards 1 nm, the variation in threshold voltage is observed to decrease from 140 mV to approximately 6 mV for NMOS transistors and from 130 to 11 mV for PMOS transistors, both of which are indications of diminishing SCE. The ON current is also observed to be independent of the channel length for small radius transistors due to the influence of large lateral electric field forcing carriers in the inversion region to travel with saturation drift velocity. Threshold voltage roll-off, DIBL, and subthreshold slope of silicon nanowire NMOS and PMOS transistors are measured and compared with earlier studies. Transient circuit performance, power dissipation and layout area of an inverter, 2-input and 3-input NAND, NOR, XOR gates and full adder circuits are measured and analyzed. As a specific case, simulation results show that the worst-case delay of a full adder circuit is 8.5 ps at no load and it increases by 0.15 ps/aF. The worst-case power dissipation of the same circuit is 23.6 nW at no load and increases approximately by 4.04 nW/aF. The layout area of the full adder is also measured to be 0.11 μm^2 which is 5 times smaller than a 6-transistor SRAM cell laid out using a 65 nm technology node. Compared to the results reported previously for silicon bulk and double-gated SOI transistors, this study indicates the silicon nanowire technology may be a potential choice for the future of VLSI circuits because of its low power dissipation in a compact layout area.

References

1. Collier CP, Wong EW, Belohradsky M, Raymo FM, Stoddard JF, Kuekes PJ, Williams RS, Heath JR (1999) Electronically configurable molecular-based logic gates. Science 285:391–394
2. Goldstein SC, Budiu M (2001) Nanofabrics: spatial computing using molecular electronics. Proc 28th Annu Int Symp Comp Archit: 178–189
3. Kim K, Das KK, Joshi RV, Chuang C-T (2005) Leakage power analysis of 25-nm double-gate CMOS devices and circuits. IEEE Trans Electron Devices 52(5):980–986
4. Tsutsui G, Saitoh M, Nagumo T, Hiramoto T (2005) Impact of SOI thickness fluctuation on threshold voltage variation in ultrathin body SOI MOSFETs. IEEE Trans Nanotechnol 4(3):369–373
5. Zhang R, Roy K, Janes DB (2001) Double-gate fully-depleted SOI transistors for low-power high performance nano-scale circuit design. Proc Int Symp Low Power Electron Design: 213–218
6. Reddy GV, Kumar MJ (2005) A new dual-material double-gate (DMDG) nanoscale SOI MOSFET-two dimensional analytical modeling and simulation. IEEE Trans Nanotechnol 4(2):260–268
7. Kumar A, Kedzierski J, Laux SE (2005) Quantum-based simulation analysis of scaling in ultrathin body device structures. IEEE Trans Electron Devices 52(4):614–617
8. Vasileska D, Ahmed SS (2005) Narrow-Width SOI devices: the role of quantum-mechanical size quantization effect and unintentional doping on device operation. IEEE Trans Electron Devices 52(2):227–236
9. Yang J-W, Fossum J (2005) On the feasibility of nanoscale triple-gate CMOS transistors. IEEE Trans Electron Devices 52(6):1159–1164
10. Yu B, Chang L, Ahmed S, Wang H, Bell S, Yang C-Y, Tabery C, Ho C, Xiang Q, King T-J, Bokor J, Hu C, Lin M-R, Kyser D (2002) FinFET scaling to 10 nm length. Tech Dig IEDM: 251–254
11. Choi Y-K, Chang L, Ranade P, Lee J-S, Ha D, Balasubramanian S, Agarwal A, Ameen M, King T-J, Bokor J (2002) FinFET process refinements for improved mobility and gate work function engineering. Tech Dig IEDM: 259–262
12. Wagner RS, Ellis WC (1964) Vapor–liquid–Solid mechanism of single crystal growth. Appl Phys Lett 4(5):89–90
13. Kamins TI, Williams SR, Basile DP, Hesjedal T, Harris JS (2001) Ti-catalyzed Si nanowires by chemical vapor deposition: Microscopy and growth mechanisms. J Appl Phys 89(2):1008–1016
14. Islam MS, Sharma S, Kamins TI, Williams RS (2004) Ultra-high-density silicon nanobridges formed between two vertical silicon surfaces. Nanotechnology 15:L5–L8
15. Kikuchi T, Moriya S, Nakatsuka Y, Matsuoka H, Nakazato K, Nishida A, Chakihara H, Matsuoka M, Moniwa M (2004) A new vertically stacked poly-Si MOSFET for 533 MHz high speed 64 mbit SRAM. Tech Dig IEDM: 923–926
16. Takato H, Sunouchi K, Okabe N, Nitayama A, Hieda K, Horiguchi F, Masuoka F (1991) Impact of surrounding gate transistor (SGT) for ultra-high-density LSI. IEEE Trans Electron Devices 38(3):573–578
17. Zheng Y, Rivas C, Lake R, Alam K, Boykin TB, Klimeck G (2005) Electronic properties of silicon nanowires. IEEE Trans Electron Devices 52(6):1097–1103
18. Miyano S, Hirose M, Masuoka F (1992) Numerical analysis of a cylindrical thin-pillar transistor (CYNTHIA). IEEE Trans Electron Devices 39(8):1876–1881
19. Dwyer C, Vicci L, Poulton J, Erie D, Superfine R, Washburn S, Taylor RM (2004) The design of DNA self-assembled computing circuitry. IEEE Trans VLSI Syst 12(11):1214–1220
20. Dwyer C, Vicci L, Taylor RM (2003) Performance simulation of nanoscale silicon rod field-effect transistor logic. IEEE Trans Nanotechnol 2(2):69–74
21. Sze SM (1981) Physics of semiconductor devices, 2nd edn. Wiley, New York

22. Choi Y-K, Ha D, King T-J, Bokor J (2003) Investigation of gate induced drain leakage (GIDL) current in thin body devices: single-gate ultrathin body, symmetrical double gate, and asymmetrical double gate MOSFETs. Jpn J Appl Phys 42:2073–2076
23. Semenov O, Pradzynski A, Sachdev M (2002) Impact of gate induced drain leakage on overall leakage of submicrometer CMOS VLSI circuits. IEEE Trans Sem Manufac 15(1):9–18
24. Wettstein A, Schenk A, Fichtner W (2001) Quantum device-simulation with density gradient model on unstructured grids. IEEE Trans Electron Devices 48(2):279–283
25. Lombardi C, Manzini S, Saporito A, Vanzi M (1988) A physically based mobility model for numerical simulation of nonplanar devices. IEEE Trans Comput Aided Design Integr Circuits Syst 7(11):1164–1171
26. Caughey DM, Thomas RE (1967) Carrier mobilities in silicon empirically related to doping and field. Proc IEEE 55(no. 12):2192–2193
27. Arora ND, Hauser JR, Roulston DJ (1982) Electron and hole mobilities in silicon as a function of concentration and temperature. IEEE Trans Electron Devices ED-29:292–295
28. Serberherr S (1984) Process and device modeling for VLSI. Microelec Reliab 24(2):225–257
29. Liu Z-H, Hu C, Huang J-H, Chan T-Y, Jeng M-C, Ko PK, Cheng YC (1993) Threshold voltage model for deep-submicrometer mosfets. IEEE Trans Electron Devices 40(1):86–95
30. Sery G, Borkar S, De V (2002) Life is CMOS: why chase the life after. Proc Design Autom Conf: 78–83
31. Chiang TK (2005) New current–voltage model for surrounding-gate metal-oxide-semiconductor field effect transistors. Jpn J Appl Phys 44(9A):6446–6451
32. Chen Q, Harrell EM, Meindl JD (2003) A physical short-channel threshold voltage model for undoped symmetric double-gate MOSFETs. IEEE Trans Electron Devices 50(7):1631–1637
33. Boeuf F et al (2004) A conventional 45 nm CMOS node low-cost platform for general purpose and low power applications. Tech Dig IEDM: 425–428
34. Numata T et al (2004) Performance enhancement of partially and fully-depleted strained-SOI MOSFETs and characterization of strained Si device parameters. Tech Dig IEDM: 177–180
35. Shima A, Ashihara H, Hiraiwa A, Mine T, Goto Y (2005) Ultra-shallow junction formation by self-limiting LTP and its application to sub-65 nm node MOSFETs. IEEE Trans Electron Devices 52(6):1165–1171
36. Wang HCH et al (2004) Low power device technology with SiGe channel, HfSiON, and Poly-Si gate. Tech Dig IEDM: 161–164
37. Luo Z et al (2004) High performance and low power transistors integrated in 65 nm bulk CMOS technology. Tech Dig IEDM: 661–664
38. Lindert N, Choi YK, Chang L, Anderson E, Lee WC (2001) Quasiplanar NMOS FinFETs with sub-100 nm gate lengths. Proc Device Res Conf: 26–27
39. Choi YK, Lindert N, Xuan P, Tang S, Ha D, Anderson E (2001) Sub-20 nm CMOS FinFET technologies. Tech Dig IEDM: 421–424
40. Wakabayashi H et al (2004) Transport properties of sub-10 nm planar-bulk-CMOS devices. Tech Dig IEDM: 429–432
41. Bai P et al (2004) A 65 nm logic technology featuring 35 nm gate lengths, enhanced strain, 8 Cu interconnect layers, low-k ILD and 0.57 μm SRAM cell. Tech Dig IEDM: 657–660
42. Kedzierski J, Dried DM, Nowak EJ, Kanarsky T, Rankin JH (2001) High performance symmetric-gate and CMOS compatible Vt asymmetric-gate FinFET devices. Tech Dig IEDM: 437–440
43. Yang FL, Chen HY, Chen FC, Chan YL (2001) 35 nm CMOS FinFETs. Proc Symp VLSI Tech: 104–105
44. Concannon A, Piccinini F, Mathewson A, Lombardi C (1995) The numerical simulation of substrate and gate currents in MOS and EPROMS. Tech Dig IEDM: 289–292
45. Dort MJV, Woerlee PH, Walker AJ (1994) A simple model for quantisation effects in heavily-doped silicon MOSFETs at inversion conditions. Solid State Electron 37(3):411–414

46. Srivastava N, Banerjee K (2004) A comparative scaling analysis of metallic and carbon nanotube interconnections for nanometer scale VLSI technologies. Proc 21st Int Multilevel Interconnect Conf: 393–398
47. Moon P, Dubin V, Johnston S, Leu J, Raol K, Wu C (2003) Process roadmap and challenges for metal barriers. Tech Dig IEDM: 841–844
48. Tada M et al (2003) A 65 nm-Node, Cu interconnect technology using porous SiOCH film (k = 2.5) covered with ultra-thin, low-k pore seal (k = 2.7). Tech Dig IEDM: 845–848
49. Nakai S et al (2003) A 65 nm CMOS technology with high performance and low-leakage transistor, a 6T-SRAM cell and robust hybrid-ULK/Cu interconnects for mobile multimedia applications. Tech Dig IEDM: 285–288
50. Kondo S, Yoon BU, Tokitoh S, Misawa K, Sone S, Shin HJ, Ohashi N, Kobayashi N (2003) Low-pressure CMP for 300-mm ultra low-k (k = 1.6 1.8)/Cu integration. Tech Dig IEDM: 151–154
51. Isobayashi A, Enomoto Y, Yamada H, Takahashi S, Kadomura S (2004) Thermally robust Cu interconnects with Cu–Ag alloy for sub 45 nm node. Tech Dig IEDM: 953–956
52. Alers GB, Sukamto J, Park S, Harm G, Reid J (2006) Containing the finite size effect in copper lines. Semicond Int 29(5):38–42
53. Alioto M, Palumbo G (2002) Analysis and comparison on full adder block in submicron technology. IEEE Trans VLSI Syst 10(6):806–823
54. Chang CH, Gu J, Zhang M (2005) A review of 0.18-μm full adder performances for tree structured arithmetic circuits. IEEE Trans VLSI Syst 13(6):686–695
55. Shams AM, Darwish TK, Bayoumi MA (2002) Performance analysis of low-power 1-Bit CMOS full adder cells. IEEE Trans VLSI Syst 10(1):20–29
56. Sayed M, Badawy W (2002) Performance analysis of single-bit full adder cells using 0.18, 0.25 and 0.35 μm CMOS technologies. IEEE Int Symp Circuits Sys: 559–562
57. Chang CH, Zhang M, Gu J (2003) A novel low power low voltage full adder cell. Proc 3rd Int Symp Image Signal Process Anal: 454–458
58. Khatibzade AA, Raahemifar K (2004) A study and comparison of full adder cells based on the standard static CMOS logic. Proc Can Conf Electr Comp Eng: 2139–2142
59. Goel S, Gollamudi S, Kumar A, Bayoumi M (2004) On the design of low-energy hybrid CMOS 1-Bit full adder cells. Proc 47th Midwest Symp Circuits Syst: 209–212

Chapter 2
Single Work Function Silicon Nanowire MOS Transistors

2.1 Device Design

2.1.1 Purpose

In the first chapter of this book, SNTs with dual work function gates were designed and their device characteristics were examined. Later in the same chapter, basic digital CMOS gates were built; their circuit performance, power dissipation, and layout characteristics were analyzed; basic SNT processing steps were shown. A dual work function CMOS technology requires the use of different metals in NMOS and PMOS transistor gates. Finding the appropriate metals that match exactly to the work function values found in this study may often be a difficult enterprise as it may require alloys for gate material or incompatible metals with the SNT processing.

One way to reduce the set of problems associated with dual metals is to use a single metal gate. Therefore, this chapter is dedicated to design NMOS and PMOS transistors with a single work function metal gate, and furthermore use these transistors in designing CMOS circuits.

As in Chapter 1, this chapter will also introduce the design criteria for SNTs that use a single metal gate, show the design flow to select optimum device dimensions for both NMOS and PMOS transistors, and analyze speed, power dissipation and layout area characteristics of various CMOS logic gates and mega cells that use these devices.

The design criteria for the single work function NMOS and PMOS SNTs are very similar to what has been applied to the dual work function SNTs.

1. NMOS and PMOS transistors need to have at least 300 mV threshold voltage for 1 V CMOS circuit operation
2. The static OFF current has to be under 1 pA in either NMOS or PMOS transistor

A. Bindal, S. Hamedi-Hagh, *Silicon Nanowire Transistors*,
DOI 10.1007/978-3-319-27177-4_2

2.1.2 The Criteria for Low Static Power Dissipation

There are three major components that produce low static power dissipation in MOS transistors.

1. Junction leakage
2. Subthreshold leakage
3. Gate-Induced-Drain-Leakage (GIDL) current

The elements that reduce the OFF current are decreasing DIBL, tox, body doping concentration, and E_S as mentioned earlier in Chapter 1. In this study, tox is set to a minimum value to maintain negligible gate leakage compared to I_{OFF}. The body doping concentration is at the intrinsic level to minimize C_D. E_S is also kept at minimum because gate-drain region does not have any overlap.

2.1.3 Device Structure

As with the dual work function SNTs, this study also considers both NMOS and PMOS transistors enhancement type with undoped silicon bodies constructed perpendicular to the substrate. Both transistors have the same body radius and effective channel length. Source/drain (S/D) contacts are assumed highly doped to obtain ohmic contacts. Both NMOS and PMOS transistors have metal gates and 1.5 nm thick gate oxide.

Device simulations are performed using Silvaco's 3-D ATLAS device simulation environment with a 1 V power supply voltage. The device radius is changed from 1 nm to 25 nm while its effective channel length is varied between 5 nm and 250 nm.

2.1.4 Physical Models Used in Device Simulations

Nano-scale devices require quantum equations to obtain accurate estimations of carrier transport in the channel region. However, ATLAS simulator has limitations of using Schrodinger's equation in full capacity to calculate effective mass and mobility values, and therefore, it follows a semiclassical approach in which the semiconductor surface potential and density of states are corrected using the density gradient method. The parameters of this method are first calibrated according to the results of self-consistent Poisson–Schrodinger equation at a negligible current flow, and then used in drift–diffusion and hydrodynamic equations to compute current densities with Fermi–Dirac carrier statistics.

For low electric field effects, Lombardi's vertical and horizontal electric field dependent mobility model is used; for high electric fields, velocity saturation and other high electric field effects are estimated by Caughey's drift velocity model. Arora's model is used for mobility degradation due to lattice temperature. To estimate the recombination rates in the bulk and at the silicon/oxide interface, concentration-dependent Shockley–Read–Hall recombination and surface recombination models are used, respectively.

Serberherr's impact ionization model constitutes the only generation model [1]. The semiconductor band-to-band tunneling mechanism producing GIDL current is not included in the simulations because of three factors. The first factor is the absence of a gate–drain (gate–source) overlap region in the cylindrical device structure: only the fringing component of the transverse electric field emanating from the edge of the gate may induce GIDL. The second factor is the decrease in transverse electric field (perpendicular to the current transport axis) compared to a bulk device with a single gate: surface band bending in bulk or a partially depleted SOI device is appreciable to promote GIDL current generation [2]. The third factor is the magnitude of the power supply voltage: the drain-to-gate potential being less than the silicon band gap is not an effective method to create enough band bending at the semiconductor surface to allow valence band electrons to tunnel into the conduction band. The gate oxide tunneling mechanisms and hot carrier injection are also ignored because these mechanisms largely depend on oxide growth and composition. Kim et al. [3] also pointed out that gate current constituted only a small percentage of the total OFF current for double-gated SOI devices with 1.5 nm gate oxide thickness.

2.1.5 Determining a Single Metal Gate Work Function

The first task of the design process is to determine a common metal gate work function that works for both NMOS and PMOS transistors as shown in Fig. 2.1. In this figure, threshold voltage, V_T, was measured as a function of metal gate work function for body radius values ranging between 1 nm and 25 nm at the minimum effective channel length of 5 nm. The intersection of NMOS and PMOS threshold voltages is projected to the x-axis to produce a common metal gate work function for each wire radius which results in a single threshold voltage for both transistors.

2.1.6 The OFF Current Requirement for the Design

The OFF current is an important factor towards lowering the standby power consumption in the entire chip. In this study, both NMOS and PMOS transistors are designed to have I_{OFF} in the proximity of 1 pA as mentioned previously. This value is significantly smaller than the OFF currents found in double-gated SOI

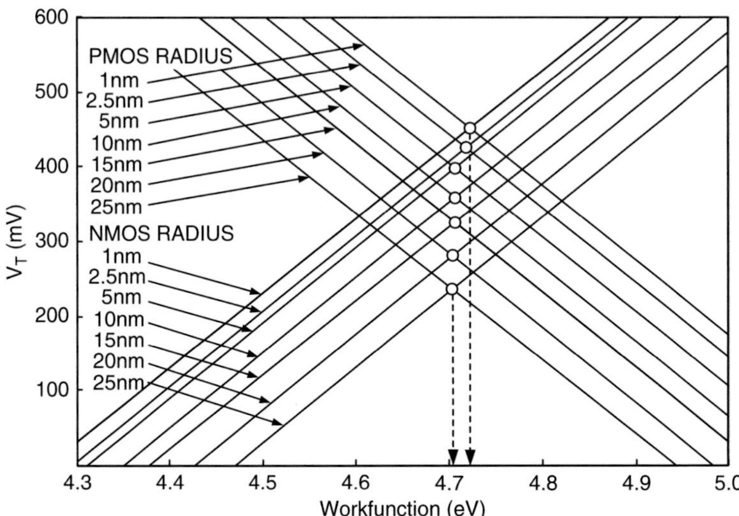

Fig. 2.1 Threshold voltages of NMOS and PMOS nanowire transistors as a function of metal work function at a minimum channel effective length of 5 nm. Both transistor radii are changed between 1 and 25 nm to produce single work function values approximately between 4.7 and 4.72 eV

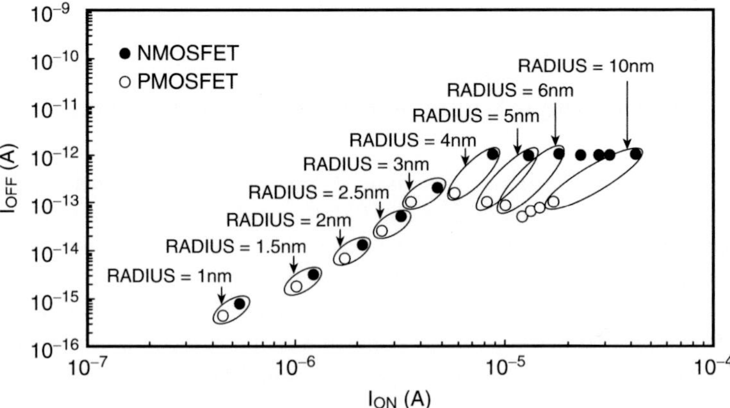

Fig. 2.2 ON versus OFF currents of NMOS and PMOS nanowire transistors. The radius of both transistors was changed between 1 and 10 nm while the effective lengths were varied between 5 and 37 nm

transistors in earlier modeling studies [3–5] and several orders of magnitude smaller than the value predicted by Sery [6]. Figure 2.2 shows I_{OFF} as a function of I_{ON} for wire radius between 1 nm and 10 nm and effective channel lengths between 5 nm and 37 nm for transistors producing 1 pA or smaller OFF currents. Table 2.1 lists the dimensions of the "selected" NMOS and PMOS transistors in Fig. 2.2.

2.1.7 Transistor Transient Characteristics: Intrinsic Transient Time

Following the device selection process for low static power dissipation, the intrinsic transient time, τ, of each transistor in Table 2.1 was measured. As described earlier in Chapter 1, the intrinsic transient time determines the time interval for a transistor to charge/discharge the gate capacitance of an identical transistor and it is plotted as a function of maximum DC transconductance, gmsat, in Fig. 2.3. The objective of this figure is to determine a common body dimension that produces maximum gmsat and minimum τ for both NMOS and PMOS transistors. In this figure, the only NMOS and PMOS device pair that exhibits the highest gmsat and the lowest intrinsic gate delay is the one with at 4 nm radius and 7 nm effective channel length, which is considered the optimum body geometry. The corresponding intrinsic transient time values of 0.5 ps for the NMOS and 0.7 ps for the PMOS transistor are also in close proximity to the results of Yu et al. [7] and Sery et al. [6].

Table 2.1 The dimensions of NMOS and PMOS nanowire transistors that satisfy IOFF < 1 pA

Body radius (nm)	Leff (nm)	Body radius (nm)	Left (nm)
1.0	5.0	5.0	12.0
1.5	5.0	6.0	16.0
2.0	5.0	7.0	22.0
2.5	5.0	8.0	27.0
3.0	5.0	9.0	32.0
4.0	7.0	10.0	37.0

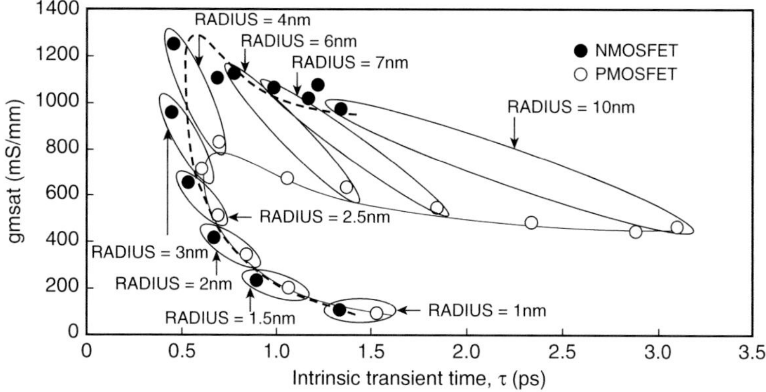

Fig. 2.3 Maximum DC transconductance versus intrinsic gate delay of the "qualified" NMOS and PMOS nanowire transistors whose leakage currents are below 1 pA

2.1.8 DC Characteristics of the Selected NMOS and PMOS Transistors

The threshold voltage roll-off of the 4 nm radius transistors with effective channel lengths ranging between 7 nm and 150 nm is measured to be 18 mV for the NMOS and 36 mV for the PMOS transistor. These values are an order of magnitude smaller than ΔV_T values of the 20 nm gate length bulk silicon transistors reported by Boeuf [8] and others [9–12].

The amount of DIBL is 114 mV/V for the NMOS and 69 mV/V for the PMOS transistors with 4 nm radius and 7 nm effective channel length. Figure 2.4 shows these values along with previously published data for comparison purposes [3, 7, 9, 11, 13, 14].

The subthreshold slope is found to be 65 mV/dec and 70 mV/dec for both NMOS and PMOS transistors at the drain voltages of 50 mV and 1 V, respectively. These results are plotted in Fig. 2.5 and show close-to-ideal characteristics in comparison with the modeling results of double-gated SOI transistors published in [3] and in the previous experimental data [7, 11–18].

Figure 2.6 shows the inverter transfer function produced by the 4 nm radius and 7 nm effective channel length NMOS and PMOS SNTs. The inverter threshold voltage (the projection of the output voltage at 0.5 V to the *x*-axis) is skewed towards 0 V due to the higher NMOS drive current; but, the inverter still produces sufficient low and high noise margins at 405 mV and 520 mV, respectively, for safe circuit operation.

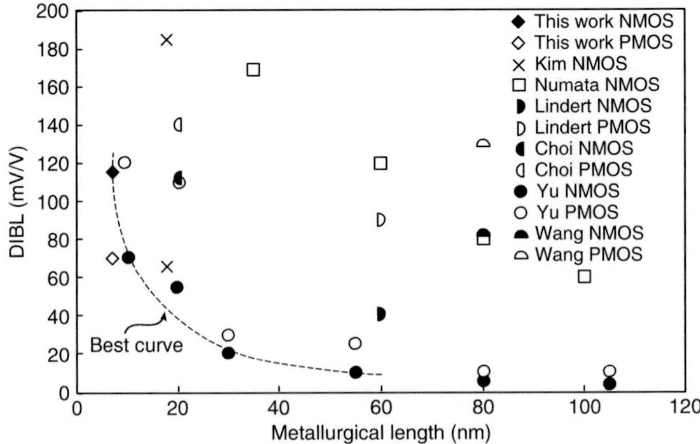

Fig. 2.4 Drain Induced Barrier Lowering (DIBL) of undoped, single work function NMOS and PMOS nanowire transistors with 4 nm body radius and 7 nm effective channel length. Prior work is included for comparison

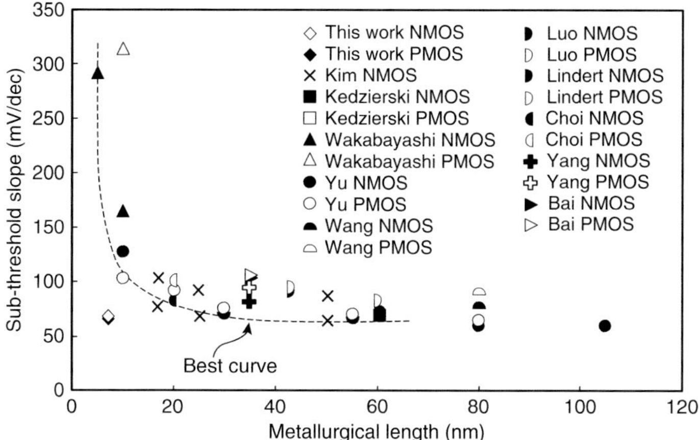

Fig. 2.5 Subthreshold slope of undoped, single work function NMOS and PMOS nanowire transistors with 4 nm body radius and 7 nm effective channel length. Prior work is included for comparison

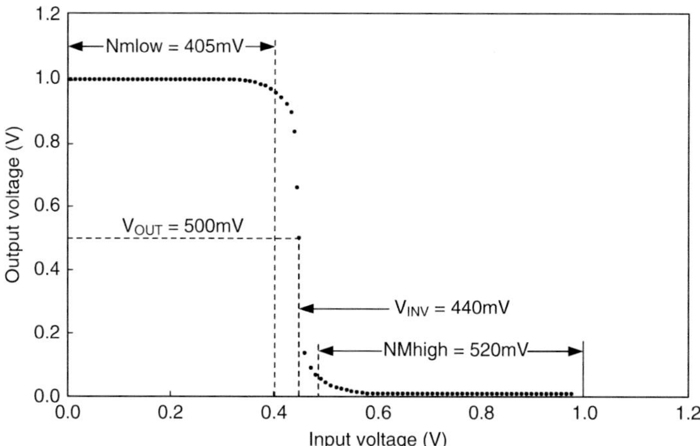

Fig. 2.6 Transfer function of an inverter built with 7 nm effective channel length and 4 nm radius NMOS and PMOS nanowire transistors. The inverter threshold voltage, V_{INV}, corresponds to the projection of Vout = 0.5 V to the x-axis

2.2 Circuit Performance

2.2.1 Parasitic Extraction and Post-layout Issues

As in the dual work function SNT study, primitive gates including an inverter, 2-input and 3-input NAND, NOR and XOR gates, and a full adder were built to measure the transient characteristics, power dissipation, and layout area of each gate. All measurements were conducted before and after parasitic layout extraction and compared with each other to understand the effects of parasitic wire resistance, capacitance, and contact resistance on circuit performance. Since these transistors are constructed perpendicular to the substrate, the minimum exposed transistor feature on the layout is 4 nm wire radius to make contacts. Copper wires with 6.4 nm width and 1.4 aspect ratio (wire height to width) are used for interconnects and 2.4 nm by 2.4 nm vias are used for contacts. Since sub-10 nm range copper wire electrical characteristics do not exist in the literature, copper resistivity value was again determined from Srivastava's model [19] as shown in Fig. 1.14 in Chapter 1. Subsequently, 20 μΩ-cm resistivity was used to calculate the sheet resistance of 6.4 nm wide metal interconnects. Similarly, contact resistance was extrapolated from the experimental data on 100 nm and larger via diameters and resulted in 18.5 Ω for metal contacts as shown in Fig. 1.15 in Chapter 1.

A simple RC calculation on inverter rise and fall times reveals approximately 25 kΩ PMOS SNT channel resistance and 11.8 kΩ NMOS SNT channel resistance. If one limits the total contact and wire resistance values to be 10 % of the equivalent n-channel resistance (or 1180 Ω) to avoid interconnect-related delays, then the discharge path can accommodate a 349 nm long copper wire between two contacts. The maximum wire length in the inverter layout is only 54 nm long. Because the equivalent PMOS channel resistance is 25 kΩ instead of 11.8 kΩ, the charge path in the inverter can even support more wire resistance. More complex circuits containing multiple transistors in series can tolerate higher numbers of contacts and longer wire lengths in charge and discharge paths. However, long wire lengths interconnecting different cells may exhibit high resistance values and produce slow nodes and larger overall circuit delays. Fortunately, in the upper-metal routing, design rules are more relaxed, allowing wider and thicker wires.

Area, fringe, and coupling capacitances of metal 1 and metal 2 wires per unit length are calculated using Ansoft's two-dimensional electrostatic solver. These capacitance values are used to extract metal-to-metal and metal-to-substrate parasitic capacitances from layouts for circuit simulations.

2.2.2 Transient Performance

The worst-case transient time and delay are shown in Figs. 2.7 and 2.8 as a function of load capacitance after layout parasitic extraction for different CMOS gates. In each figure, a capacitance value of 40 aF corresponds to a fan-out of ten SNTs, considering the gate capacitance of a single transistor is approximately 4 aF.

The worst-case transient times shown in Fig. 2.7 are essentially the worst-case rise times since a PMOS transistor has higher equivalent channel resistance compared to an NMOS transistor. The worst-case transient times of the inverter, 2-input and 3-input NAND gates overlap with each other primarily due to the single PMOS transistor charging the output capacitance. The worst-case transient times of the 2-input NOR and XOR circuits cluster together because two PMOS transistors in

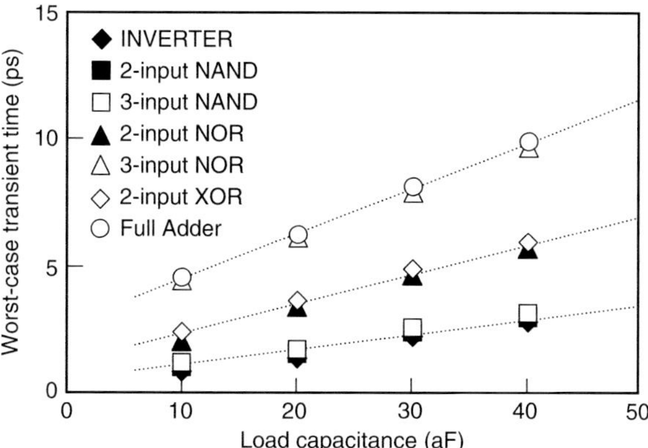

Fig. 2.7 Post-layout worst-case transient time characteristics of various primitive gates built with 7 nm effective channel length and 4 nm body radius NMOS and PMOS nanowire transistors

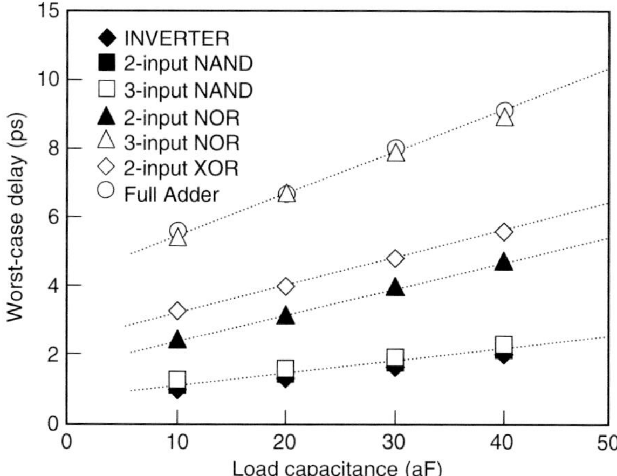

Fig. 2.8 Post-layout worst-case propagation delay characteristics of various primitive gates built with 7 nm effective channel length and 4 nm body radius NMOS and PMOS nanowire transistors

series charge the output load. The full adder and 3-input NOR circuits are in close proximity and reveal the highest worst-case transient times because the number of PMOS transistors in series increases from two to three in the critical charging path. For example, the worst-case transient times of the 2-input NAND gate and full adder in Fig. 2.7 are expressed as $T = 0.378 + 0.052C_L$ and $T = 3.21 + 0.179C_L$ in picoseconds, respectively, where C_L is the output capacitance in aF.

As in the dual work function SNTs, Fig. 2.8 shows similar characteristics compared to the worst-case transient times in Fig. 2.7 because each CMOS logic gate in this figure uses the same critical charging and discharging paths to compute worst-case delays. For example, the worst-case delays of the 2-input NAND gate and the full adder circuits are expressed as $T_D = 0.667 + 0.033C_L$ and $T_D = 4.45 + 0.124C_L$ in picoseconds, respectively, where C_L is the output capacitance in aF. Worst-case gate delay values obtained from CMOS circuits that use SNTs are significantly smaller in comparison with the CMOS circuits that use bulk silicon or SOI technologies. Inverter gate delays of 4 ns and 5 ns from a chain of double-gated SOI and bulk silicon inverters [3] are substantially larger compared to the 1 ps inverter gate delay obtained in this study.

The effect of gate layout parasitics on transient performance is substantial when there is no capacitive load and decreases proportionally as the output capacitance increases. Worst-case post-layout gate delays at no capacitive load increase between 17 % and 36 % after layout extraction. This change primarily stems from the layout complexity, transistor count and number of series transistors on the critical path. For example, the worst-case delay of the full adder increases by 36 % after parasitic extraction when there is no output load and decreases to 14.7 % for a fan-out of six transistors.

2.2.3 *Dynamic Power Dissipation*

The worst-case dynamic power dissipation of various CMOS gates is shown in Fig. 2.9 as a function of frequency, f, when a 10 aF capacitive load is connected to each logic gate's output. Worst-case power dissipation is obtained by considering all the possible input combinations to a logic gate, measuring the average value of the power supply current within one clock period (activity factor = 1 %) for each combination, and finally selecting the combination that yields the maximum average current. Each current waveform is averaged within one clock period during charging and discharging cycle of the output capacitance. In general, worst-case power dissipation increases with increasing transistor count, layout complexity, and the number of "parallel" charging or discharging paths to a capacitive load. For example, the worst-case power dissipations of the 2-input NAND gate and full adder circuits are expressed as $P = 0.33 + 32.97f$ and $P = 0.66 + 59.93f$ in nanowatts, respectively, in Fig. 2.9.

Figure 2.10 shows the worst-case power dissipation figures of each CMOS gate as a function of load capacitance at 1 GHz. For example, the worst-case

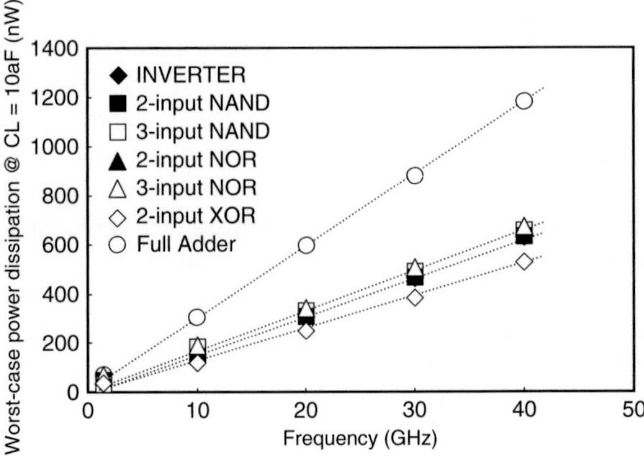

Fig. 2.9 Post-layout worst-case power dissipation of various primitive gates built with 7 nm effective channel length and 4 nm body radius NMOS and PMOS nanowire transistors at a capacitive load of 10 aF

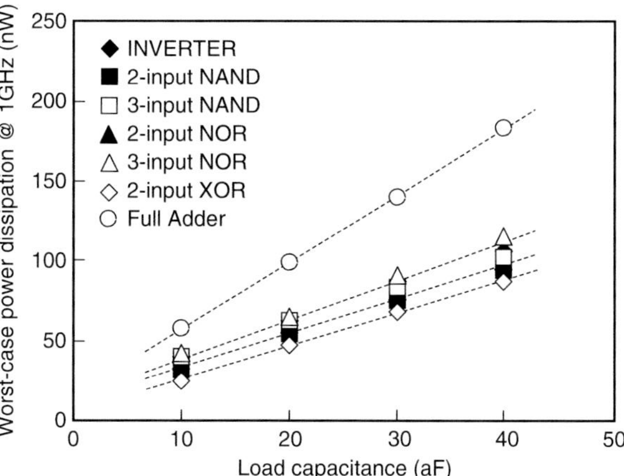

Fig. 2.10 Post-layout worst-case power dissipation of various primitive gates built with 7 nm effective channel length and 4 nm body radius NMOS and PMOS nanowire transistors at 1 GHz

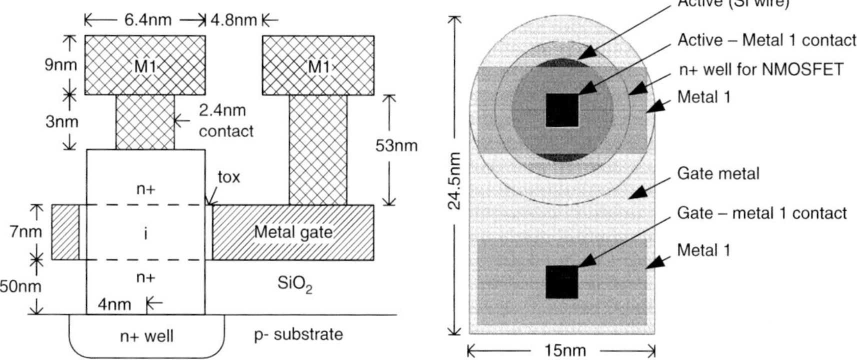

Fig. 2.11 Cross section and layout of a single, 7 nm effective channel length and 4 nm body radius NMOS transistor. Note that an N-well (shown by dashed lines in the cross-sectional view) surrounds only the p^+ well of PMOS transistors

power dissipations of the 2-input NAND gate and full adder circuits produce $P = 12.67 + 2.06C_L$ and $P = 19.44 + 4.11C_L$ in nanowatts, respectively.

2.2.4 Cell Layout Area Estimations

An inverter, 2-input and 3-input NAND, NOR and XOR circuits, and a full adder were laid out using the 4 nm radius nanowire transistors. Figure 2.11 shows the cross section and the corresponding layout of a single SNT. The active region defines the circular body of the device which is surrounded by an N-well if the transistor is an n-channel device or a P-well if it is a p-channel. The outmost circle represents the metal gate and is connected with a rectangular gate extension. All

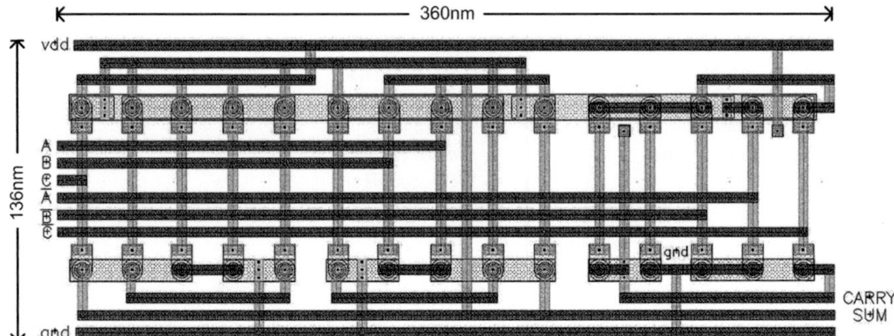

Fig. 2.12 Full adder layout using 7 nm effective channel length and 4 nm body radius NMOS and PMOS nanowire transistors. A, B, and C are the two inputs of the full adder and the carry-in, respectively. $\overline{A}, \overline{B}$, and $\overline{C}$ correspond to the two complemented inputs of the full adder and the carry-in, respectively

Table 2.2 Layout area of various primitive gates built with 7 nm effective channel length and 4 nm body radius NMOS and PMOS nanowire transistors

Gates	Area (nm^2)
Inverter	5712
2-input NAND	6624
3-input NAND	8736
2-input NOR	6624
3-input NOR	8736
2-input XOR	10,752
Full Adder	48,960

contacts are indicated by 2.4 nm by 2.4 nm black squares touching the drain (source) and the gate of the transistor.

Figure 2.12 shows the layout of a full adder. The vertical dimension is fixed at 136 nm in all cell layouts. Each circular structure corresponds to a vertical NMOS or PMOS transistor. Interconnections are established by 6.4 nm wide metal 1 and metal 2 wires. Power and ground connections are made to the P- and N-wells with multiple contacts, a metal 1 layer, a via and a metal 2 layer. The P-well is completely surrounded by an N-well to prevent latch-up. Layout areas of the primitive gates used in this study are listed in Table 2.2, which are considerably smaller than the state-of-the-art counterparts. The 28-transistor full adder in this study has a cell area of approximately 0.049 μm^2. Extrapolation of a 28-transistor CMOS full adder layout area in 350 nm [20], 180 nm [21], and 45 nm technology nodes towards the 7 nm technology node still produces a layout area of approximately 40 μm^2, which is more than 800 times larger than the full adder area shown in Fig. 2.12.

Table 2.3 Circuit performance, power dissipation, and layout area of 28-transistor full adder in this study and earlier work

Lg (nm)	V_{DD}(V)	f_{op} (MHz)	P_T (nW)	Delay (ps)	Area (μm²)	References
350	3.3	[a]	164,000	227	[a]	[54]
350	1.2	50	2490	2037	387	[54]
350	1.8	50	6090	827	387	[54]
350	2.5	50	12,820	528	387	[54]
350	3.3	50	24,120	406	387	[54]
350	3.3	[a]	65,000	400	[a]	[54]
250	3.3	[a]	58,000	300	[a]	[54]
180	3.3	[a]	30,000	100	[a]	[54]
180	1.0	100	2500	650	[a]	[54]
180	1.8	100	6230	292	100	[54]
180	1.0	100	1450	756	100	[54]
180	1.8	300	345	195	[a]	[54]
180	1.8	50	11	327	[a]	[54]
7	1.0	1000	118[b]	7.5	0.05	[54]

[a]Cases not reported
[b]An output load of 24 aF (6 transistor gates)

2.2.5 *Full Adder Comparison*

In order to acknowledge the significance of this technology with respect to the earlier and emerging technologies, a 28-transistor CMOS full adder circuit is examined in terms of transient performance, power dissipation, and layout area in various technology nodes. The results of this study are tabulated in Table 2.3, which shows the full adder in this study excels in all three categories [20–26].

2.3 Summary

Three-dimensional undoped NMOS and PMOS SNTs with a single-work-function metal gate were designed to minimize the leakage current under 1 pA and maximize the DC transconductance as a function of device radius and effective channel length. Device simulations were performed in Silvaco's Atlas device design environment to produce transistor DC characteristics such as ON and OFF currents, ΔV_T, DIBL, and S. Transient performance, power dissipation, and layout area of an inverter, multi-input NAND, NOR and XOR gates, and full adder circuits were measured and analyzed. As a specific case, simulation results showed that the worst-case transient time and the worst-case delay for the 2-input NAND gate are 1.63 ps and 1.46 ps, respectively, and for the full adder 7.51 ps and 7.43 ps, respectively. The worst-case power dissipation is 62.1 nW for the two-input

NAND gate and 118.1 nW for a full adder operating at 1 GHz for the same output capacitance. The layout areas are 0.0066 μm^2 for the 2-input NAND gate and 0.049 μm^2 for the full adder circuits. Compared to the results previously reported on silicon bulk and double-gated SOI transistors, these data indicate the silicon wire technology is a potential choice for the future of VLSI circuits because of overall low gate delay and transient times, compact layout area, and low static and dynamic power dissipation.

References

1. Serberherr S (1984) Process and device modeling for VLSI. Microelectron Reliab 24:225–257
2. Choi Y, Ha D, King T, Bokor J (2003) Investigation of gate induced drain leakage (GIDL) current in thin body devices: single-gate ultra-thin body, symmetrical double gate, and asymmetrical double gate MOSFETs. Japan J Appl Phys 42:2073–2076
3. Kim K, Das K, Joshi R, Chuang C (2005) Leakage power analysis of 25-nm double gate CMOS devices and circuits. IEEE Trans Electron Devices 52:980–986
4. Zhang R, Roy K, Janes D (2001) Double-gate fully-depleted SOI transistors for low-power high performance nano-scale circuit design. ISLPED. pp 213–218
5. Yang J, Fossum J (2005) On the feasibility of nanoscale triple-gate CMOS transistors. IEEE Trans Electron Devices 52:1159–1164
6. Sery G, Borkar S, De V (2002) Life is CMOS: why chase the life after? DAC. pp 78–83
7. Yu B (2002) FinFET scaling to 10 nm length. IEDM. pp 251–254
8. Boeuf F (2004) A conventional 45 nm CMOS node low-cost platform for general purpose and low power applications. IEDM. pp 425–428
9. Numata T (2004) Performance enhancement of partially and fully-depleted strained-SOI MOSFETs and characterization of strained-Si device parameters. IEDM. pp 177–180
10. Shima A, Ashihara H, Hiraiwa A, Mine T, Goto Y (2005) Ultrashallow junction formation by self-limiting LTP and its application to sub-65 nm node MOSFETs. IEEE Trans Electron Devices 52:1165–1171
11. Wang H (2004) Low power device technology with SiGe channel, HfSiON, and poly-Si gate. IEDM. pp 161–164
12. Luo Z (2004) High performance and low power transistors integrated in 65 nm bulk CMOS technology. IEDM. pp 661–664
13. Lindert N, Choi Y, Chang L, Anderson E, Lee W (2001) Quasi-planar NMOS FinFETs with sub-100 nm gate lengths. Device Research Conf. pp 26–27
14. Choi Y, Lindert N, Xuan P, Tang S, Ha D, Anderson E (2001) Sub-20 nm CMOS FinFET technologies. IEDM. pp 421–424
15. Wakabayashi H (2004) Transport properties of sub-10 nm planar-bulk-CMOS devices. IEDM. pp 429–432
16. Bai P (2004) A 65 nm logic technology featuring 35 nm gate lengths, enhanced strain, 8 Cu interconnect layers, low-k ILD and 0.57 μm2 SRAM cell. IEDM. pp 657–660
17. Kedzierski J, Dried D, Nowak E, Kanarsky T, Rankin J (2001) High performance symmetric-gate and CMOS compatible Vt asymmetric-gate FinFET devices. IEDM. pp 437–440
18. Yang F, Chen H, Chen F, Chan Y (2001) 35nm CMOS FinFETs. Symp. VLSI Technol. pp 104–105
19. Srivastava N, Banerjee K (2004) A comparative scaling analysis of metallic and carbon nanotube interconnections for nanometer scale VLSI technologies. Proc. 21st Int. Multilevel Interconnect Conf. pp 393–398

20. Alioto M, Palumbo G (2002) Analysis and comparison on full adder block in submicron technology. IEEE Trans VLSI Syst 10:806–823
21. Chang C, Gu J, Zhang M (2005) A review of 0.18 μm full adder performances for tree structured arithmetic circuits. IEEE Trans VLSI Syst 13:686–695
22. Chang C, Zhang M, Gu J (2003) A novel low power low voltage full adder cell. Proc. 3rd Int. Symp. Image Signal Proc. Analysis. pp 454–458
23. Shams A, Darwish T, Bayoumi M (2002) Performance analysis of low-power 1-bit CMOS full adder cells. IEEE Trans VLSI Syst 10:20–29
24. Sayed M, Badawy W (2002) Performance analysis of single-bit full adder cells using 0.18, 0.25 and 0.35 μm CMOS technologies. IEEE Int. Symp. Circuits Syst. pp 559–562
25. Khatibzade A, Raahemifar K (2004) A study and comparison of full adder cells based on the standard static CMOS logic. Canadian Conf. Electron. and Comput. Eng. pp 2139–2142
26. Goel S, Gollamudi S, Kumar A, Bayoumi M (2004) On the design of low-energy hybrid CMOS 1-bit full adder cells. 47th Midwest Symp. Circuits and Syst. pp 209–211

Chapter 3
SPICE Modeling for Analog and Digital Applications

3.1 BSIMSOI Device Parameters

3.1.1 Introduction

In Chapters 1 and 2, we studied the device and digital circuit aspects of dual and single work function SNTs with the intention to minimize power dissipation [1–3]. Both of these studies have determined that silicon nanowire technology is better suited for the future of VLSI in terms of circuit speed and power dissipation compared to dual-gated SOI or FiNFET technologies [4, 5]. These studies also included the weaknesses of SNTs such as increased layout area due to surrounding gate metal thickness, large source resistance caused by source contact extension, and limited ON current caused by fixed transistor geometry.

In Chapters 1 and 2, SPICE level 6 models were used in circuit simulations [6, 7]. While these models had acceptable accuracy in producing circuit speed and power dissipation figures for basic CMOS logic gates, more accurate intrinsic device modeling and parasitic RC extraction were required for simulating larger scale digital circuits, analog circuits, and Radio Frequency (RF) circuits. This need prompted us to explore more accurate SPICE models such as BSIMSOI for fully depleted Silicon-On-Insulator (SOI) devices [8] to use in the circuit simulations.

Even though various analytical physical models were proposed by different research groups for nanowire transistors, these results were only satisfactory for long channel devices; large discrepancies arose between the simulated and modeled I–V characteristics in sub-10 nm channel length range [9–12]. The inaccuracies were especially intolerable in the active device region of operation where impact ionization and DIBL are dominant factors in determining the transistor linearity for analog and RF applications.

Therefore, the main focus of this chapter is to obtain accurate BSIMSOI models by curve-fitting the BSIMSOI model parameters to the I–V curves resulted from ATLAS device simulations. Calculation of voltage dependent intrinsic gate oxide

A. Bindal, S. Hamedi-Hagh, *Silicon Nanowire Transistors*,
DOI 10.1007/978-3-319-27177-4_3

capacitance, device parasitic resistors and capacitors including the high frequency effective gate resistance is also presented in this chapter. In addition, transistor input and output I–V characteristics, transconductance and output resistance at various biasing conditions, extraction of S-parameters, power gains, and computation of f_{max} and f_T are part of this study. To reduce ohmic losses, the basic SNT layout used in earlier chapters was revised and multiple contacts were used at the drain and source terminals of the transistor.

3.1.2 The Device

The device structure of the vertical SNT was shown in Fig. 1.1 of Chapter 1. However, a more detailed representation of this device is illustrated in Fig. 3.1. The NMOS transistor is built perpendicular to an N-well (PMOS on P-well) SOI substrate. Both NMOS and PMOS transistors are constructed as enhancement type with undoped silicon bodies with high source and drain doping concentrations to achieve ohmic contacts. Source and drain junctions are extended from both ends of the channel by 12 nm.

In Chapter 1, NMOS and PMOS transistors with 40 nm channel length and 4 nm radius used two kinds of gate metals and produced 300 mV threshold voltage, and neither allowed more than $I_{OFF} = 1$ pA. Both of these SNTs had almost equal I_{ON} and intrinsic transient times. However, the dynamic power consumption was not considered as one of the critical design parameters in determining the body dimensions of either device.

In this chapter, we will add the interplay between the intrinsic energy and intrinsic transient time on top of the existing design requirements, the 300 mV threshold voltage and $I_{OFF} < 1$ pA, in order to optimize both the dynamic power consumption and the transconductance of the device.

The first task in the design process is to determine individual gate metal work functions for NMOS and PMOS transistors to produce a threshold voltage of approximately 300 mV. Therefore, for each 10 nm channel length transistor, threshold voltage is measured and plotted as a function of gate work function for device radius between 2 nm and 20 nm as shown in Fig. 3.2. The intersection of threshold voltage with 300 mV level is projected to the x-axis to determine the gate work function of each NMOS and PMOS transistor at a different body radius.

The device selection process in this chapter is slightly different than the one given in Chapter 1. Transistor body dimensions are determined by three design objectives in sequence:

1. Select metal work functions for each NMOS and PMOS transistor to produce 300 mV threshold voltage
2. For each body radius, increase the effective device length such that the device produces 1 pA or less OFF current

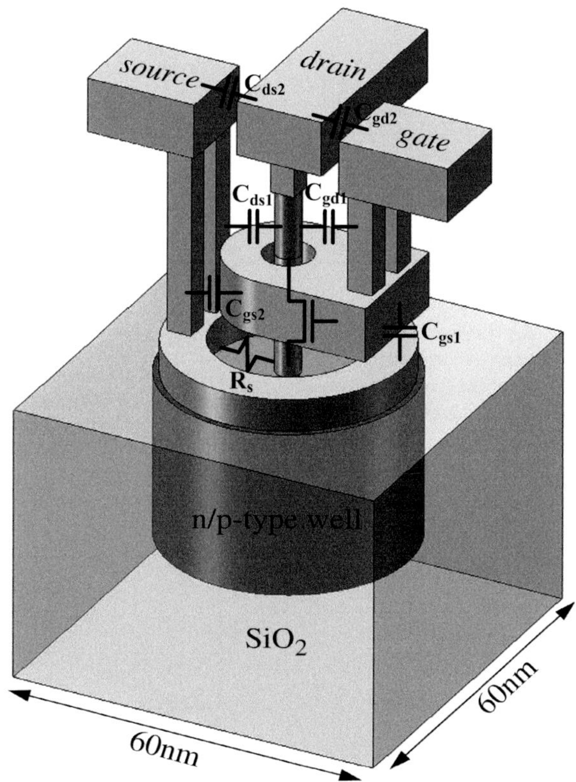

Fig. 3.1 A detailed 3D view of the SNT with major parasitic R and C components

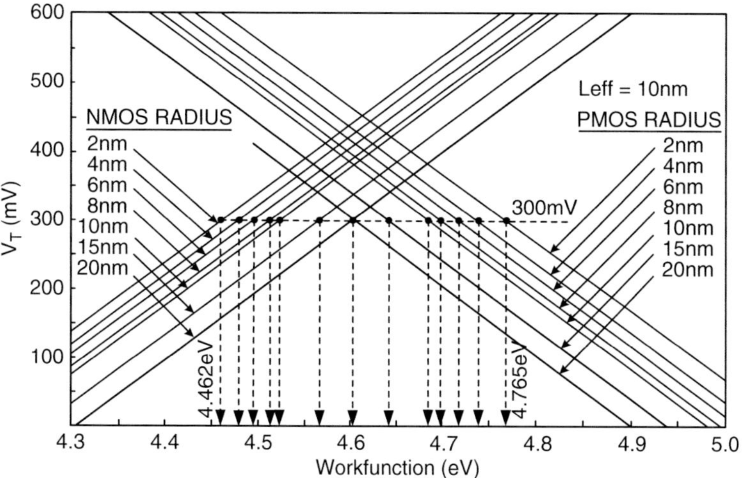

Fig. 3.2 Threshold voltage versus metal work function for $L_{EFF} = 10$ nm

Table 3.1 The SNT body dimensions that produce $I_{OFF} \leq 1$ pA

Radius (nm)	L_{EFF} (nm)
2	10
4	25
6	38
8	47
10	54
15	59
20	65

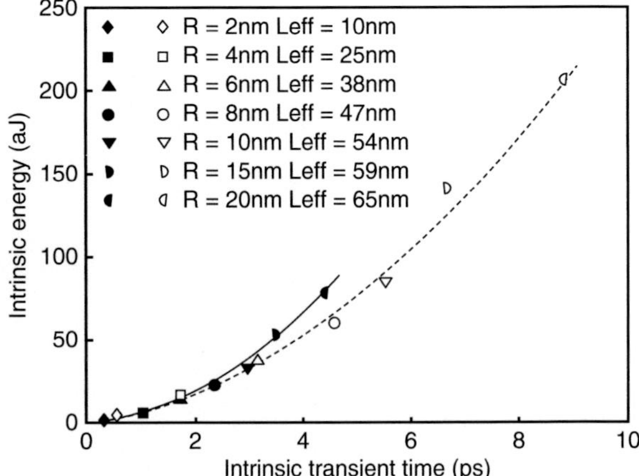

Fig. 3.3 Intrinsic energy as a function of intrinsic transient time for transistors with $I_{OFF} \leq 1$ pA

3. For each transistor satisfying the conditions in (a) and (b), determine the minimum intrinsic energy and intrinsic transient time in order to minimize the dynamic power dissipation and maximize the drive current.

Table 3.1 shows the transistor dimensions that meet the design criteria in (a) and (b). The intrinsic transient time and intrinsic energy for each NMOS and PMOS transistor in Table 3.1 are subsequently measured and plotted against each other as shown in Fig. 3.3. Here, the intrinsic transient time determines the time interval for a transistor to charge (or discharge) the gate capacitance of an identical transistor when it is fully on, and it is a quick way of measuring the speed of a transistor. Intrinsic energy, on the other hand, corresponds to the integration of instantaneous power delivered (or received) to (or from) the gate capacitance of an identical transistor as a function of time and measures the dynamic power dissipation of the transistor. Therefore, the most desirable transistor geometry in Fig. 3.3 is the one with minimal intrinsic transient time and intrinsic energy, which is 2 nm radius and 10 nm effective channel length. In our later studies when designing larger scale digital circuits, analog and RF circuits, we will primarily use the BSIMSOI models based on this device geometry in all circuit simulations.

The output I–V characteristics of the NMOS and PMOS SNTs with 2 nm radius and 10 nm channel length are shown in Figs. 3.4 and 3.5, respectively. Saturation

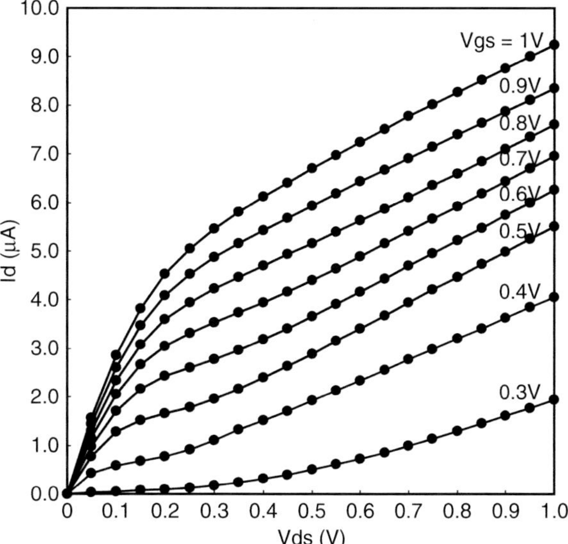

Fig. 3.4 Output I–V characteristics of a 10 nm channel length, 2 nm radius NMOS SNT obtained from numerical simulations

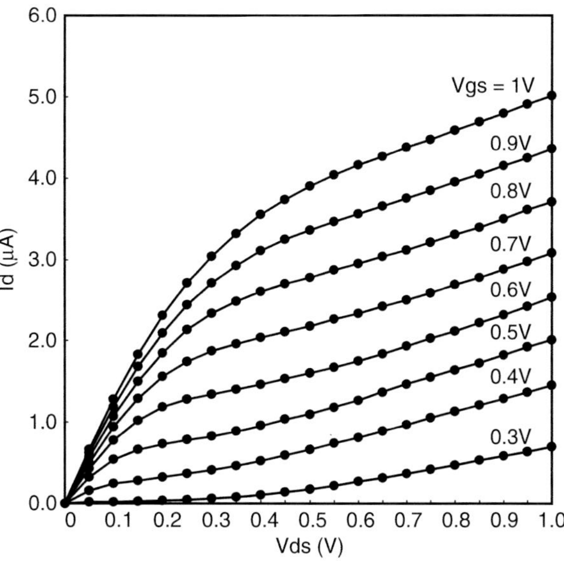

Fig. 3.5 Output I–V characteristics of a 10 nm channel length, 2 nm radius PMOS SNT obtained from numerical simulations

region does not produce a constant current in either transistor possibly because of the source-drain tunneling mechanism and the lack of a substrate contact. ON currents are approximately 9 μA for the NMOS transistor and 5 μA for the PMOS transistor.

According to the new layout rules, the minimum spacing between two metal 1 interconnects is 5 nm. The minimum width of the metal 1 interconnect is 14 nm.

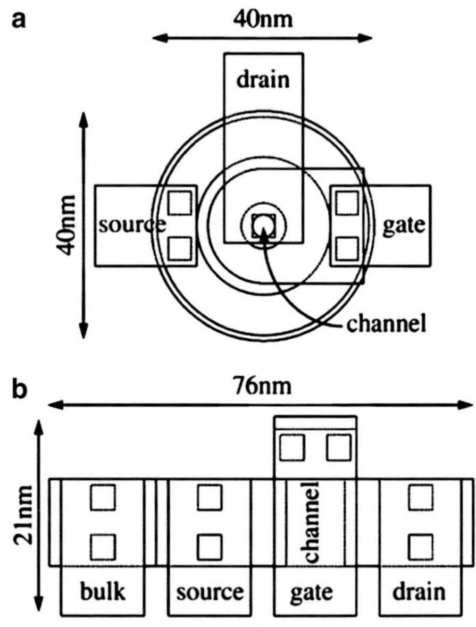

Fig. 3.6 Layout views of (**a**) SNT and (**b**) planar bulk MOSFET

Gate and source terminals have larger contact resistances due to their distance to metal 1 layer. Therefore, two parallel contacts are used for each source and gate terminals as shown in Fig. 3.6a. The N-well (P-well) is also surrounded by a concentric metal ring to reduce the overall resistance of the source terminal. The concentric source contact has 6 nm thickness, 6 nm width, and 13 nm inner radius with a 19 nm outer radius. The N-well or P-well overlap with the concentric source contact is 1 nm. The minimum side-overlap of metal 1 and contact region is 1 nm. The distance between metal 1 and metal 2 is 36 nm. Metal 2 and via 12 (connecting metal 1 and metal 2 layers) have the same design rules as metal 1 and the contact region. SNTs have negligible gate-drain overlap and drain-source parasitic capacitances compared to planar MOSFETs. SNT and the planar MOSFET layouts are shown in Fig. 3.6a, b, respectively. The layout of the planar MOSFET has the same active device and contact areas as well as metal and contact spacing and included here for comparison purposes.

Down-scaling of bulk MOSFETs has introduced various quantum mechanical effects and has increased the complexity of device modeling required for accurate circuit simulation [13]. While the channel length down-scaling of the bulk MOSFETs increases the device speed, it also introduces strong electric fields in channel causing early drain-source voltage breakdown. Therefore, reducing the channel length often requires a Lightly Doped Drain (LDD) region to ensure proper device operation. To prevent a strong vertical electric field build-up in the channel and to stop carriers to tunnel through thin oxide (causing high gate leakage current), the gate voltage of the transistor must also be reduced [14]. Continuous down-scaling of voltage levels and device dimensions of the planar MOSFETs have been proven to weaken the gate control over the channel, enhance short channel effects, and produce high OFF current, punch-through, lower carrier mobility, and gate leakage current [15].

Another major issue in the nanoscale bulk MOSFETs is the latch-up. Extra spacing between NMOS and PMOS transistors and the addition of low resistance substrate contacts are required to minimize latch-up problem at the expense of increasing the layout area [16]. When bulk MOSFETs operate at high frequencies, strong signal coupling through resistive silicon substrate increases the interaction among different circuit blocks, deteriorating signal integrity [17]. To minimize substrate coupling, guard rings and shallow trench isolations must be used despite the increase in the layout area [18]. Intrinsic undoped single-gate, double-gate, triple-gate SOI transistors or SNTs have minimal substrate coupling and latch-up issues compared to bulk MOSFETs and hence they are the preferred choices for mixed signal VLSI applications.

3.1.3 Intrinsic Modeling and Parasitic Extraction

NMOS and PMOS SNT SPICE models are created by BSIMSOI parameters which are primarily used for fully depleted SOI thin film transistors. Important SNT model parameters are listed in Table 3.2. These parameters are adjusted such that the input and output I–V transistor characteristics closely match to those obtained from numerical device simulations for 10 nm channel length and 2 nm radius SNTs. Intrinsic input I–V characteristics in the subthreshold region are shown in Fig. 3.7. Figure 3.8 shows the intrinsic input I–V characteristics and Fig. 3.9 shows the intrinsic output characteristics of SNTs for Vgs values between 0 V and 1 V. In all these three figures, small deviations between the BSIMSOI modeling and numerical simulations show the accuracy of these models for circuit simulation.

To be able to produce the parasitic RC components in Fig. 3.1, we used the transistor I–V characteristics as explained below.

The intrinsic drain-source coupling capacitance, C_{dsi}, of the fully depleted channel is obtained by Eq. 3.1.

$$C_{dsi} = \varepsilon_{Si}.\varepsilon_0 \frac{\pi R^2}{L} \tag{3.1}$$

Here, $\varepsilon_{SI} = 11.7$ is the relative permittivity of the silicon, $\varepsilon_0 = 8.854 \times 10^{-14}$ F/cm is the permittivity of free air, R is the channel radius, and L is the effective channel length. Therefore, C_{dsi} is equal to 0.13 aF. The gate oxide capacitance of the transistor, C_{ox}, is obtained by Eq. 3.2.

Table 3.2 List of important BSIMSOI intrinsic model parameters for SNTs

Parameters	Values
Channel length (L)	10 nm
Channel radius (R)	2 nm
Gate oxide thickness (t_{ox})	1.5 nm
Channel doping concentration (n_{ch})	1.5e+019 cm^{-3}
Substrate doping concentration (n_{sub})	1.0e+011 cm^{-3}
Threshold voltage (V_{th0})	0.26 V (nmos) −0.28 (pmos)
Mobility (U_0)	1000 cm^2/V.s (nmos) 300 cm^2/V.s (pmos)
Parasitic resistance per unit area (R_{dsw})	130 Ω.μm (nmos) 360 Ω.μm (nmos)
Saturation velocity (V_{sat})	≈2e+06 cm/s
Subthreshold region offset voltage (V_{off})	0.06 V
Channel length modulation (P_{clm})	25
Primary output resistance DIBL effect (P_{diblc1})	1.02e–006
Secondary output resistance DIBL effect (P_{diblc2})	1
Primary short channel effect on V_{th} (D_{vt0})	3.8
Secondary short channel effect on V_{th} (D_{vt1})	2.75
Short channel body bias effect on V_{th} (D_{vt2})	0 V^{-1}
Primary narrow width effect on V_{th} (D_{vt0w})	0
Secondary narrow width effect on V_{th} (D_{vt1w})	7.25e+007
Narrow width body bias effect on V_{th} (D_{vt2w})	0.34 V^{-1}
Subthreshold region DIBL coefficient (Eta_0)	0.008
Subthreshold body bias DIBL effect (Eta_b)	0.174 V^{-1}
DIBL coefficient exponent (D_{sub})	1
Drain/Source to channel coupling capacitance (C_{dsc})	1.373e–010 F/cm^2

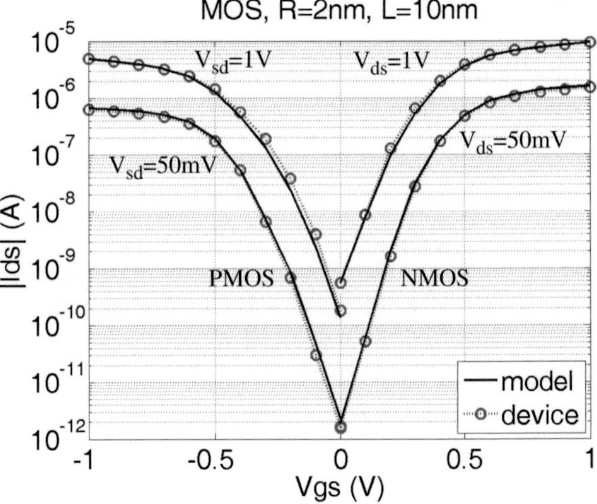

Fig. 3.7 Subthreshold input I–V characteristics of intrinsic NMOS and PMOS SNTs

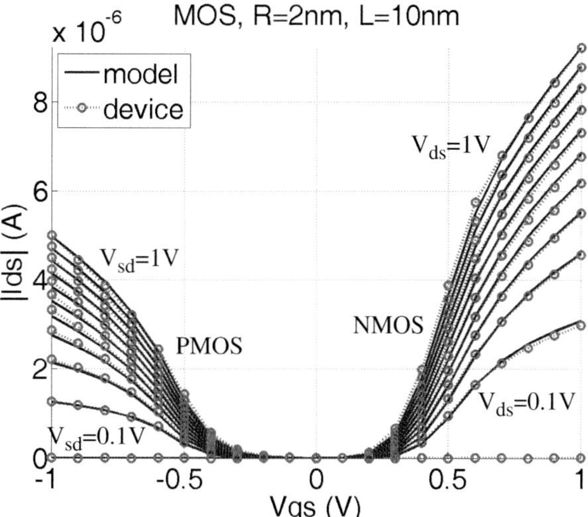

Fig. 3.8 Input I–V characteristics of intrinsic NMOS and PMOS SNTs

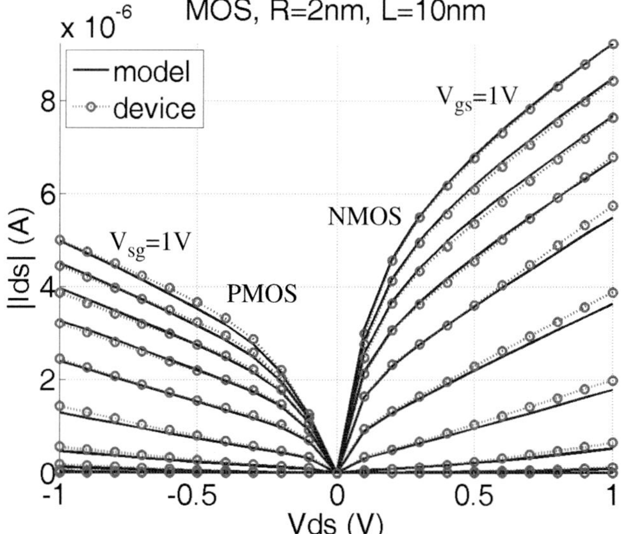

Fig. 3.9 Output I–V characteristics of intrinsic NMOS and PMOS SNTs

$$C_{ox} = \varepsilon_{Si}.\varepsilon_{ox}\frac{2\pi R}{L} \tag{3.2}$$

Here, $\varepsilon_{ox} = 3.9$ is the relative permittivity of SiO_2 the silicon oxide and tox = 1.5 nm is the oxide thickness. Thus, C_{ox} becomes equal to 2.9 aF.

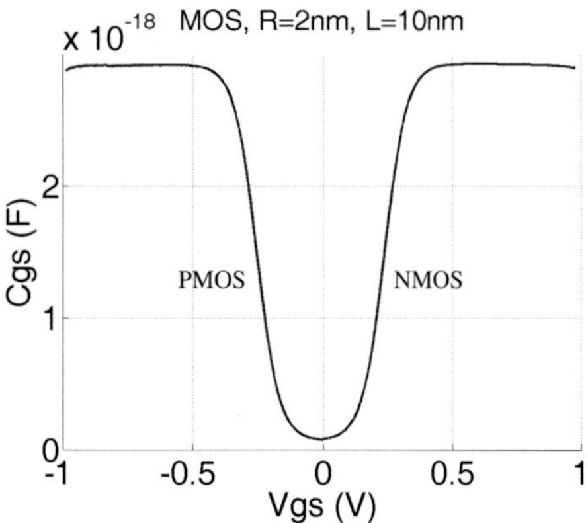

Fig. 3.10 Intrinsic gate oxide capacitance of NMOS and PMOS SNTs

The intrinsic gate-source capacitance values, C_{gs}, of the NMOS and PMOS SNTs are shown in Fig. 3.10. These capacitances are obtained from transient simulations using Eq. 3.3.

$$C_{gsi} = \frac{I(t)}{dV(t)/dI(t)} \tag{3.3}$$

Here, I(t) represents the feed current through the gate-source terminal and dV(t)/dt represents the rate of change of the charging voltage across the gate and source terminals. When the transistor turns off, the intrinsic gate-source capacitance becomes negligible due to the fully depleted channel and becomes equal to Eq. 3.4. When the transistor turns on, the channel inverts and the intrinsic gate-source capacitance reaches C_{ox}.

$$C_{gsi} = \frac{C_{ox}C_{dsi}}{C_{ox} + C_{dsi}} \tag{3.4}$$

At very small dimensions, parasitic interconnect resistance and capacitance become dominant components due to small metal thickness and close metal spacing. Therefore, parasitic values need to be accurately calculated for analog and RF circuit modeling. The parasitic RC components are modeled separately for NMOS and PMOS transistors and shown in Fig. 3.11a, b, respectively.

In these figures, C_{gsx} is the parasitic capacitance between the metal gate and the concentric source. Similarly, C_{gsy} is the parasitic capacitance between the metal gate and the source contact. The resistor, r_g, accounts for the effective gate resistance at high frequencies caused by the distributed gate-oxide-channel. The resistance, R_g, accounts for two parallel gate contacts. C_{dsx} is the parasitic capacitance between intrinsic drain and source contacts, and C_{dsy} is the parasitic capacitance between drain and source interconnects. The resistors, R_{sx} and R_{sy}, represent source contacts, and the resistors, R_{nw} and R_{pw}, represent overall concentric N-well and P-well resistances from intrinsic source to extrinsic source contacts for NMOS

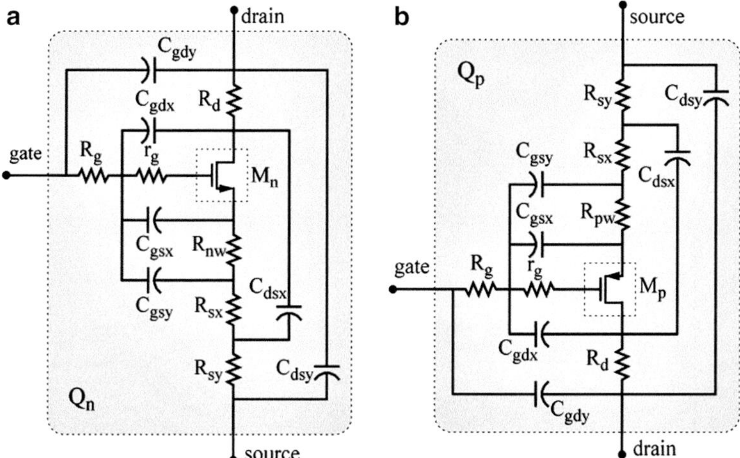

Fig. 3.11 Distributed parasitic components across (**a**) intrinsic NMOS, Mn, and (**b**) intrinsic PMOS, Mp, SNTs

and PMOS SNTs, respectively. C_{gdx} is the parasitic capacitance between the gate contact and the intrinsic drain, and C_{gdy} is the parasitic capacitance between the gate and drain interconnects. The resistor, R_d, represents the drain contact of the transistor.

All parasitic coupling and fringe capacitor values are calculated using three-dimensional ADS momentum field simulator. Besides effective resistance r_g, all other parasitic resistors are calculated by $\rho l/S$ at low frequencies where $\rho = 180\,\Omega$-nm is the resistivity of the sub-10 nm copper wire [4], l is the length of the resistive path, and S is the cross-section area of the resistive path. The resistance, r_g, affects the impedance and noise matching of the transistor at RF and is obtained by considering the high frequency gate current density distribution resulted from the distributed RC network of the gate oxide and the channel as shown in Fig. 3.12a, b. When a high frequency signal is applied to one end of the metal gate ($\theta = 0$, where θ represents the angle of the surrounding metal gate), the signal travels through two parallel signal paths (1st and 2nd) as shown in Fig. 3.12b until it reaches the other end of the metal gate ($\theta = \pm\pi$). While the high frequency current signal travels through metal gate, it leaks through distributed gate oxide capacitor. The current density of the signal decreases from the maximum value of J_m to zero as shown in Fig. 3.12a.

The effective distributed high frequency gate resistance is, therefore, given by [19].

$$r_g = \frac{1}{2}\frac{\frac{1}{\pi}\int_0^{\pi} V(\theta)d\theta}{\frac{1}{\pi}\int_0^{\pi} I(\theta)d\theta} = \frac{1}{2}\frac{\frac{1}{\pi}\int_0^{\pi}\left(R_s\frac{R\theta}{L}\right)J_cSd\theta}{\frac{1}{\pi}\int_0^{\pi} J_cSd\theta} \tag{3.5}$$

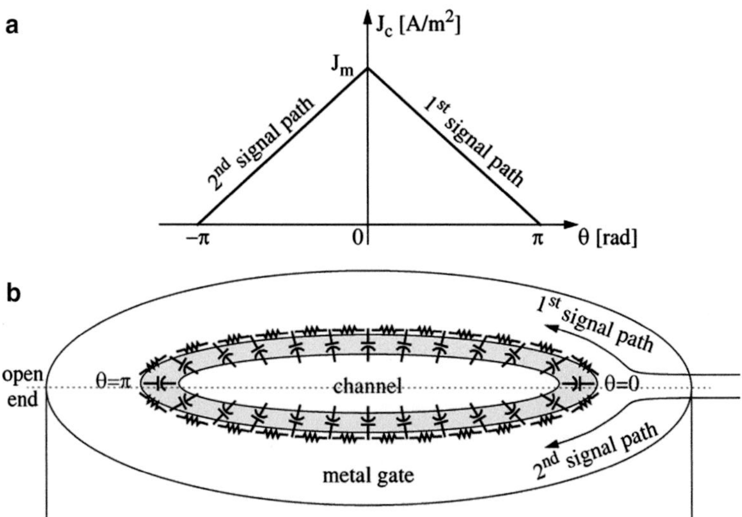

Fig. 3.12 Gate and channel (**a**) high frequency current distribution and (**b**) parasitic RC components

where R_s is sheet resistance of the metal gate, L is the channel length, S is the cross-section area of the metal gate, R is the radius of the nanowire transistor, and J_c is the current density distribution given by $J_m.(\pi\text{-}\theta)/\pi$. Here, J_m is the maximum current density in the metal gate and θ is the surrounding gate angle. The effective resistor, r_g, is therefore given by

$$r_g = \frac{1}{12}\left(R_s\frac{2\pi R}{L}\right) \tag{3.6}$$

This value is equal to the effective gate resistance of the planar transistors with signals applied to both ends of the gate. However, due to the excess polysilicon resistance of planar MOSFETs and the wide metal gate structure of SNTs, the overall effective gate resistance of SNTs is expected to be smaller than the one for planar transistors. A small rg significantly enhances the operation speed and impedance matching capabilities of the transistor.

The distributed SNT parasitic components are listed in Table 3.3a, b for resistors and capacitors, respectively. A comparison between the SNT and planar bulk MOSFET layouts reveals that SNTs suffer from larger source resistance and gate-source capacitance while MOSFETs suffer from larger source/drain junctions, gate-drain and gate-source capacitances.

Table 3.3 List of SNT parasitic values (a) resistors and (b) capacitors

(a)		(b)	
Resistors	Values	Capacitors	Values
r_g	10 Ω	C_{gsx}	3 aF
R_g	110 Ω	C_{gsy}	1 aF
R_{nw} (R_{pw})	2.3 (3.4) k Ω	C_{gdx}	0.5 aF
R_{sx}	100 Ω	C_{gdy}	0.8 aF
R_{sy}	100 Ω	C_{dsx}	0.5 aF
R_{sy}	100 Ω	C_{dsx}	0.5 aF
R_d	70 Ω	C_{dsy}	0.8 aF

3.1.4 Extrinsic Modeling and Parasitic Extraction

The extrinsic input I–V characteristics of SNTs, including all parasitic resistances, are shown in Fig. 3.13 in 0.1 V increments of V_{ds}. The extrinsic NMOS and PMOS saturation currents are very close to the intrinsic saturation current values at 9 μA and 5 μA, respectively. The extrinsic SNT transconductances calculated from the slopes of the input I–V characteristics reach a maximum value of 20 μA/V at $V_{gs} = 500$ mV and 11.5 μA/V at $V_{sg} = 550$ mV as shown in Fig. 3.14.

The extrinsic output I–V characteristics of NMOS and PMOS SNTs are shown in Fig. 3.15. Extrinsic output conductances calculated from inverse slopes of the output I–V characteristics in Fig. 3.15 are shown in Fig. 3.16. At biasing conditions of $V_{gs} = 0.5$ V and $V_{ds} = 0.5$ V, NMOS and PMOS transistors exhibit 400 kΩ and 1.1 MΩ output resistances, respectively.

The SNT S-parameters are shown in Fig. 3.17. These parameters are obtained by sweeping the frequency from 1 MHz to 10^3 THz and using ports with $Z_0 = 1$ kΩ internal resistance to ensure stability. The input inductors, input capacitors, and the output bias-T circuits are selected to behave as open and short circuits, respectively, at sweeping frequency. SNTs are biased with $V_{ds} = 1$ V and $V_{gs} = 0.5$ V to yield maximum transconductance and high power gain. The S_{22} (output return loss) is a measure of the transistor output resistance and S_{21} (forward gain) is a measure of the transistor voltage gain. However, g_m and r_{out} of these transistors differ quite a bit from each other due to the intrinsic device characteristics. Therefore, it is expected to see S_{22} and S_{21} of the NMOS and PMOS transistors deviate from each other while S_{11} (input return loss) and S_{12} (reverse gain) match more closely, as shown in Fig. 3.17.

The two important figures of merit for SNTs operating at RF are the maximum frequency of oscillation, f_{max}, and the unity-current-gain cutoff frequency, f_T. f_{max} is obtained when the magnitude of the maximum available power gain, G_{max}, of the transistor becomes unity and f_T is obtained when the magnitude of the current gain, H_{21}, of the transistor becomes unity. G_{max} and H_{21} of the transistor are expressed in terms of S-parameters using Eqs. 3.7 and 3.8 under simultaneous conjugate impedance matching conditions at the input and output ports.

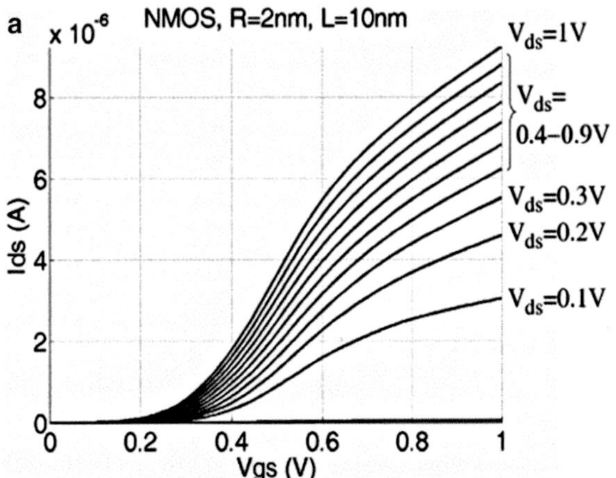

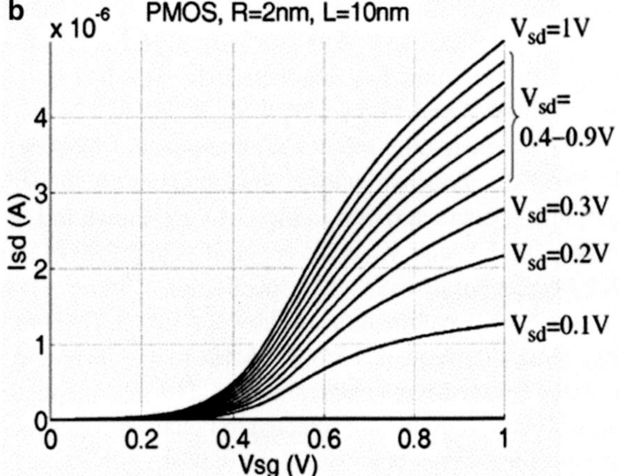

Fig. 3.13 Extrinsic input I–V characteristics of (**a**) NMOS and (**b**) PMOS SNTs

$$G_{max} = \frac{S_{21}^2}{\left(1 - S_{11}^2\right)\left(1 - S_{22}^2\right)} \tag{3.7}$$

and

$$H_{21} = \frac{S_{21}}{(1 - S_{11})(1 + S_{22}) + S_{12}S_{21}} \tag{3.8}$$

The magnitudes of G_{max} and H_{21} are plotted as a function of frequency in Fig. 3.18. From this figure, f_{max} and f_T of the NMOS SNT are 120 THz and 36 THz, respectively. Similarly, f_{max} and f_T of the PMOS SNT are 100 THz and 25 THz, respectively.

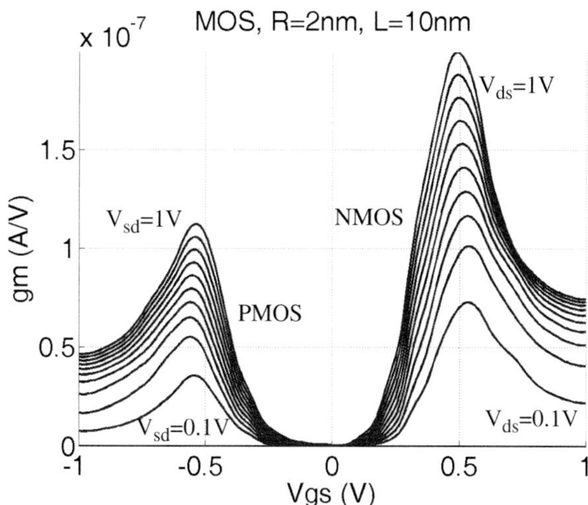

Fig. 3.14 Extrinsic transconductances of NMOS and PMOS SNTs

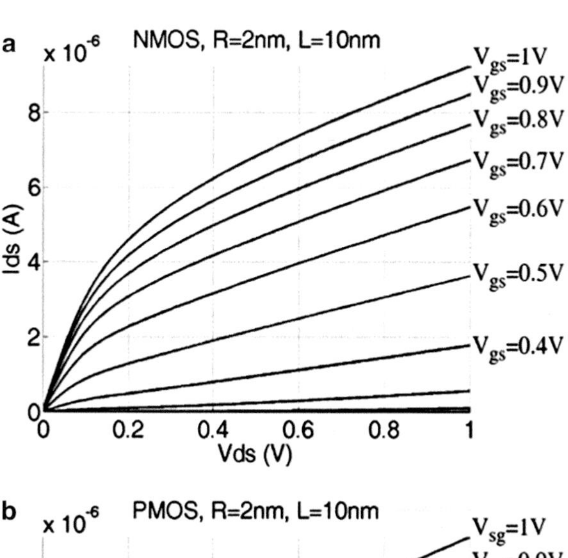

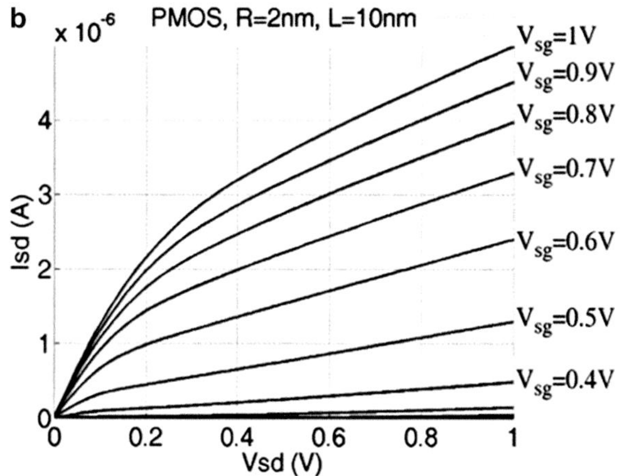

Fig. 3.15 Extrinsic output I–V characteristics of (**a**) NMOS and (**b**) PMOS SNTs

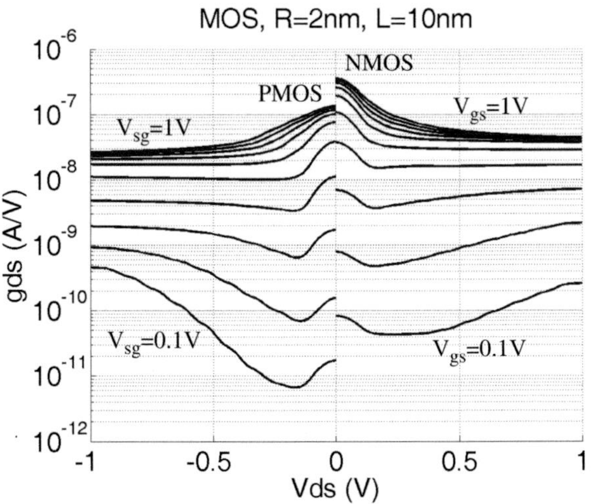

Fig. 3.16 Extrinsic output conductances of NMOS and PMOS SNTs

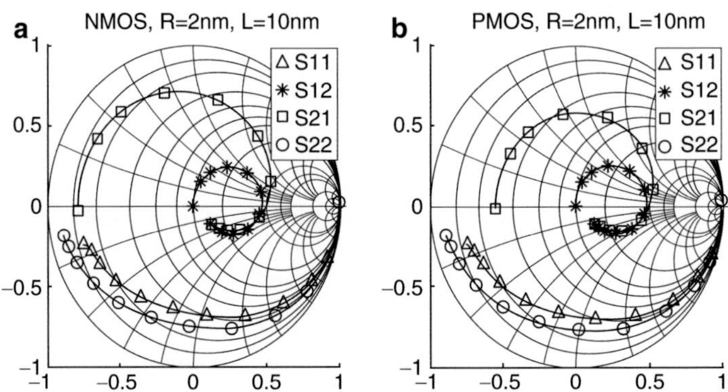

Fig. 3.17 Extrinsic S-parameters of (**a**) NMOS and (**b**) PMOS SNTs

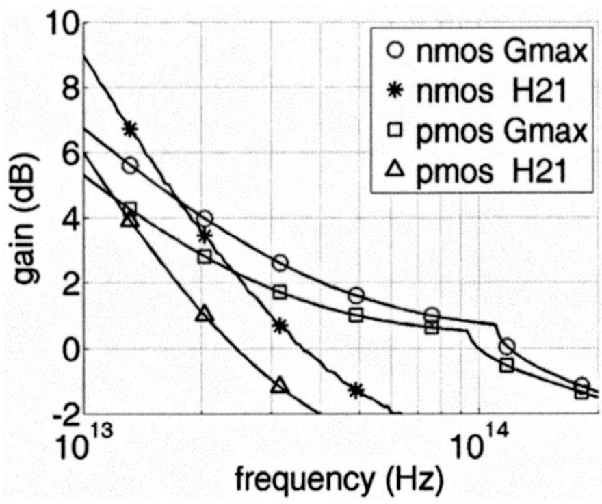

Fig. 3.18 Extrinsic power and current gains of NMOS and PMOS SNTs

3.2 Summary

To prove the values of nanoscale SNTs in digital, analog, and RF applications, there is a need to generate accurate SPICE models for transistors with channel lengths approaching 10 nm range. This chapter presents a SPICE modeling study for NMOS and PMOS SNTs with 10 nm channel length and 2 nm channel radius. The fully depleted BSIMSOI parameters are extracted from the input and output I–V characteristics of SNTs produced by three-dimensional Silvaco ATLAS device simulations. The distributed parasitic RC components of the NMOS and PMOS transistors are calculated and added to the SPICE models as subcircuits. The low frequency, high frequency, small-signal, and large-signal characteristics of the NMOS and PMOS SNTs are measured to demonstrate the capabilities of these transistors for high speed digital, analog, and RF applications.

The layout of the SNTs was optimized to achieve the highest operating frequency and gain. In analog and RF applications, ohmic losses at the device contacts as well as long source and gate extensions can deteriorate the overall gain, bandwidth, and noise characteristics of these devices. A concentric lower contact is specifically built for the source terminal for this purpose. Two additional contacts are placed to the source to further reduce the overall source resistance. Similarly, two parallel contacts are placed at the gate terminal to improve the gate resistance of these transistors.

When biased at $V_{ds} = 0.5$ V and $V_{gs} = 0.5$ V, the NMOS and PMOS SNTs exhibit 2 μA and 0.7 μA drain currents, 14 μA/V and 8 μA/V transconductances, 400 kΩ and 1.1 MΩ output resistances, 36 THz and 25 THz unity-current-gain cutoff frequencies, and 120 THz and 100 THz maximum frequency of oscillations, respectively. All these results indicate the potential use of vertical SNTs for the next-generation VLSI technologies.

References

1. Bindal A, Hamedi-Hagh S (2007) An exploratory study on power efficient silicon nanowire dynamic NMOSFET/PMESFET logic. IEE Proc Sci Measurement Technol 1:121–130
2. Bindal A, Hamedi-Hagh S (2007) Static NMOS circuits using silicon nanowire technology for crossbar architectures. Semicond Sci Tech 22:54–64
3. Bindal A, Hamedi-Hagh S (2006) The design and analysis of dynamic NMOSFET/PMESFET logic using silicon nanowire technology. Semicond Sci Tech 21:1002–1012
4. Bindal A, Naresh A, Yuan P, Nguyen KK, Hamedi-Hagh S (2007) The design of dual work function CMOS transistors and circuits using silicon nanowire technology. IEEE Trans Nanotechnol 6:291–302
5. Bindal A, Hamedi-Hagh S (2006) The impact of silicon nanowire technology on the design of single work function CMOS transistors and circuits. Nanotechnol 17:4340–4351
6. Bindal A, Hamedi-Hagh S (2007) The design of a new spiking neuron using silicon nanowire technology. Nanotechnol 18:1–12

7. Bindal A, Hamedi-Hagh S (2007) Silicon nanowire transistors and their applications for the future of VLSI: an exploratory design study of a 16 × 16 SRAM. J Nanoelectron Optoelectron 2:294–303
8. Pin Su, Fung SKH, Wyatt PW, Wan H, Mansun Chan, Niknejad AM, Hu C (2003) A unified model for partial-depletion and full-depletion SOI circuit designs: using BSIMPD as a foundation. Proceedings of the IEEE custom integrated circuits conference. pp 241–244
9. Eminente S, Alessandrini M, Fiegna C (2004) Comparative analysis of the RF and noise performance of bulk and single-gate ultra-thin SOI MOSFETs by numerical simulation. Solid State Electron 48:543–549
10. Kilchytska V, Neve A, Vancaillie L, Levacq D, Adriaensen S, Van Meer H, De Meyer K, Raynaud C, Dehan M, Raskin J, Flandre D (2003) Influence of device engineering on the analog and RF performances of SOI MOSFETs. IEEE Trans Electron Dev 50:577–588
11. Jimenez D, Iniguez B, Sune J, Saenz J (2004) Analog performance of the nanoscale double-gate metal-oxide-semiconductor field-effect-transistor near the ultimate scaling limits. J Appl Phys 96:5271–5276
12. Flandre D, Raskin J, Vanhoenacker D (2001) SOI CMOS transistors for RF microwave applications. Int J High Speed Electron Syst 11:1159–1248
13. Majima H, Saito Y, Hiramoto T (2001) Impact of quantum mechanical effects on design of nanoscale narrow channel n- and p-type MOSFETs. Technical digest of international electron device meeting. pp 951–954
14. Woerlee P, Knitel M, Van Langevelde R, Klaassen D, Tiemeijer L, Scholten A, Zegers-Van Duijnhoven A (2001) RF-CMOS performance trends. IEEE Trans Electron Dev 48:1776–1782
15. Frank D, Dennard R, Nowak E, Solomon PM, Taur Y, Wong HSP (2001) Device scaling limits for Si MOSFETs and their application dependencies. Proc IEEE 89:259–288
16. Menozzi R, Lanzoni M, Fiegna C, Sangiorgi E, Ricco B (1990) Latch-up testing in CMOS IC's. IEEE J Solid State Circ 25:1010–1014
17. Jin W, Eo Y, Shim J, Eisenstadt W, Park M, Yu H (2001) Silicon substrate coupling, noise modeling and experimental verification for mixed signal integrated circuit design. Digest of the IEEE International Microwave Symposium 3:1727–1730
18. Raskin J, Vivian A, Flandre D, Colinge J (1997) Substrate crosstalk reduction using SOI technology. IEEE Trans Electron Dev 44:2252–2261
19. Jin X (1998) An effective gate resistance model for CMOS RF and noise modeling. Technical digest of the IEEE electron device meeting. pp 961–964

Chapter 4
High-Speed Analog Applications

4.1 Introduction

In Chapter 3, we obtained accurate BSIMSOI SPICE models for NMOS and PMOS silicon nanowire transistors to replicate the simulated device I–V characteristics. We also calculated voltage-dependent intrinsic gate oxide capacitance, parasitic device resistors and capacitors, including the high frequency effective gate resistance in order to generate accurate extrinsic circuit models for SNTs. From these extrinsic models, we reproduced the realistic input and output I–V characteristics, transconductance, and output resistance curves as a function of biasing conditions, S-parameters, power gains, f_{max}, and f_T. This chapter presents the first application of silicon nanowire technology on analog circuits. Various small and large-signal analog circuits such as a single-stage CMOS amplifier, a differential pair amplifier, and a two-stage operational amplifier were designed and simulated using the BSIMSOI SPICE models of SNTs.

4.2 Brief Description of Transistor Design and Modeling

The optimal SNT body dimensions of 2 nm radius and 10 nm channel length are determined according to the minimum intrinsic transient times and minimum static and dynamic power dissipations for each NMOS and PMOS transistor discussed in Chapter 3. The BSIMSOI device models are based on these particular device dimensions and they are used for all the circuit simulations in this chapter.

A. Bindal, S. Hamedi-Hagh, *Silicon Nanowire Transistors*,
DOI 10.1007/978-3-319-27177-4_4

4.3 Single-Stage CMOS SNT Amplifier

4.3.1 The CMOS Amplifier Design

The layout of an SNT CMOS amplifier which can be used as a large-signal inverter or a small-signal amplifier is shown in Fig. 4.1a. The biasing point of transistors is decided by the voltage level of the input signal. For 1 V supply voltage, both transistors operate in active region when the input voltage changes between 0.4 V and 0.6 V. Two PMOS transistors, Qp1 and Qp2, are used in parallel to ensure their overall ON current is almost equal to that of a single NMOS transistor, Qn. The area of the SNT amplifier is 70×144 nm. The gate and source contacts for each transistor in this layout are doubled to reduce ohmic losses at these terminals which, otherwise, act as a degenerative factor to decrease the voltage gain of the amplifier.

The low frequency small-signal model of the amplifier, when both transistors operate in active region, is shown in Fig. 4.1b. In this figure, R_d represents the cumulative drain resistance and R_s is the cumulative source resistance given by

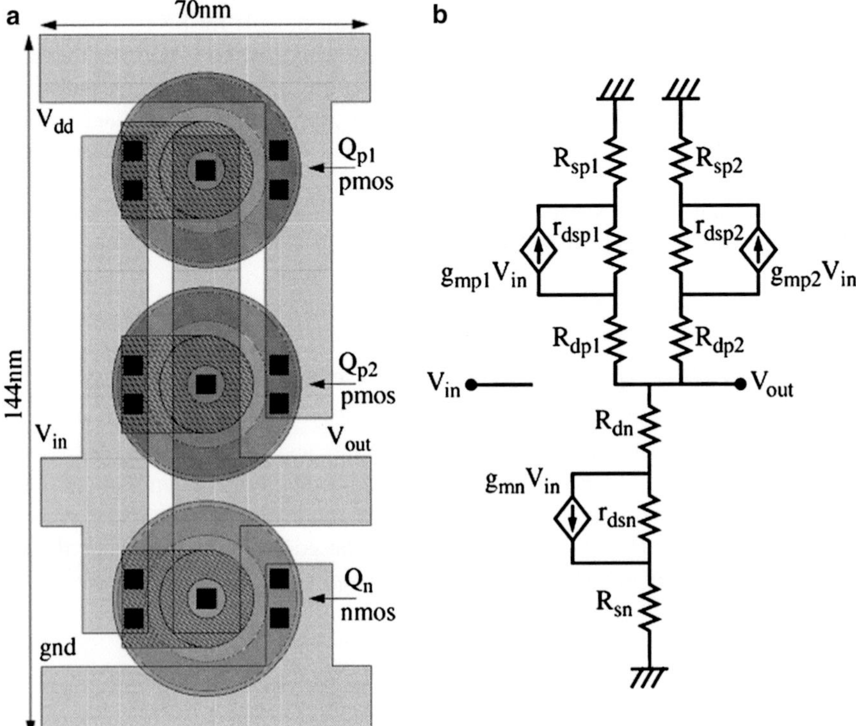

Fig. 4.1 The SNT CMOS amplifier (**a**) layout and (**b**) low frequency small-signal model

$R_{sn} = R_{sx} + R_{sy} + R_{nw}$ for the NMOS SNT and $R_{sp} = R_{sx} + R_{sy} + R_{pw}$ for the PMOS SNT. The actual values are given in Chapter 3.

According to Fig. 4.1b, the low frequency small-signal gain of the CMOS amplifier is given by Eq. 4.1.

$$A_{v0} = -\left[\frac{g_{mn} r_{dsn}}{R_n} + \frac{g_{mp1} r_{dsp1}}{R_{p1}} + \frac{g_{mp2} r_{dsp2}}{R_{p2}}\right]\left(R_n // R_{p1} // R_{p2}\right) \tag{4.1}$$

where

$$R_n = r_{dsn} + R_{sn} + R_{dn} \approx r_{dsn} \tag{4.2}$$

and

$$R_{p1} = r_{dsp1} + R_{sp1} + R_{dp1} \approx r_{dsp1} \tag{4.3}$$

and

$$R_{p2} = r_{dsp2} + R_{sp2} + R_{dp2} \approx r_{dsp2} \tag{4.4}$$

Therefore, Eq. 4.1 can be rewritten as

$$A_{v0} \approx -\left[g_{mn} + g_{mp1} + g_{mp2}\right]\left(r_{dsn} // r_{dsp1} // r_{dsp2}\right) \tag{4.5}$$

Here, A_{V0} is approximately equal to -6.9 with $g_{mn} = 14$ μA/V and $g_{mp1} = g_{mp2} = 8$ μA/V, $r_{dsn} = 400$ kΩ and $r_{dsp1} = r_{dsp2} = 1.1$ MΩ at $V_{ds} = 0.5$ V and $V_{gs} = 0.5$ V. The simplified low frequency voltage gain of the SNT CMOS amplifier becomes similar to the MOSFET amplifier gain when $R_s \ll 1/g_m \ll r_{ds}$ is satisfied.

4.3.2 *The Characteristics of the CMOS Amplifier*

The frequency bandwidth and linearity are the two important figures of merit for an amplifier. When biased at $V_{gs} = 0.5$ V, the maximum gain of the CMOS amplifier becomes $20 \times \log(6.5) = 17$ dB and the phase becomes 180° with a unity voltage gain cutoff frequency of 20 THz at a phase angle of 60° as shown in Fig. 4.2. The 3 dB frequency bandwidth of the amplifier is approximately 500 GHz and the power dissipation becomes 1.64 μW. The output spectrum of the amplifier in response to a two-tone test with 10 GHz spacing is shown in Fig. 4.3. The 2nd and 3rd harmonic distortions, HD2 and HD3 tones, of the amplifier are 10 dB and 20 dB below the

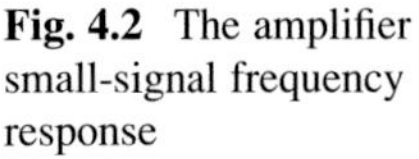

Fig. 4.2 The amplifier small-signal frequency response

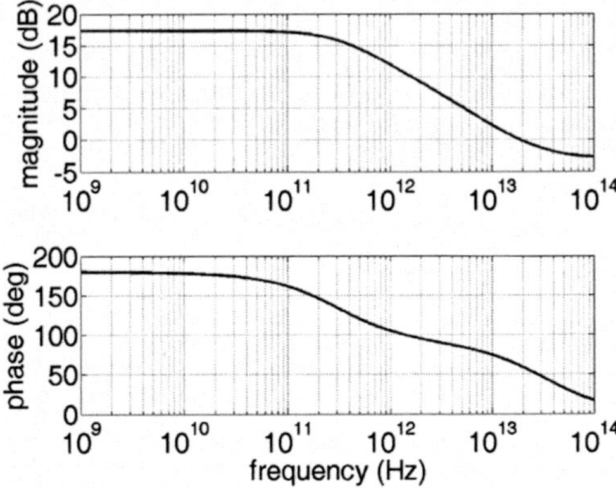

Fig. 4.3 The amplifier two-tone output spectrum

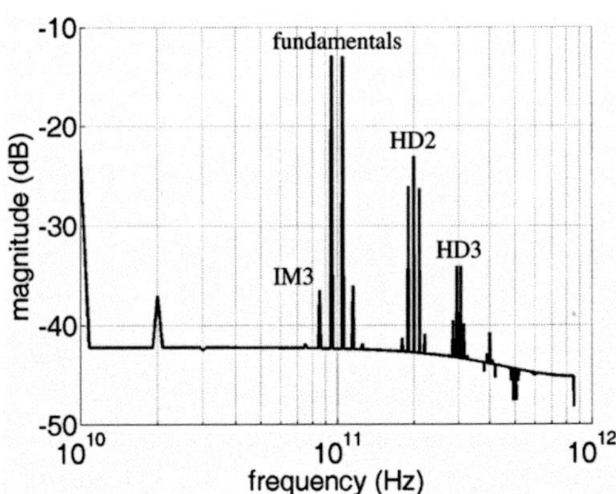

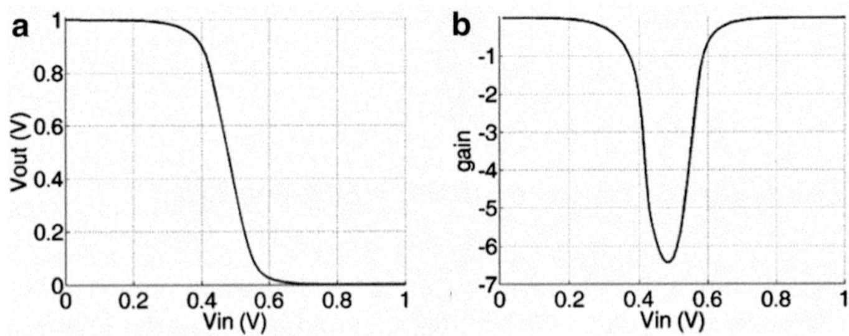

Fig. 4.4 The amplifier large-signal (**a**) transfer characteristic and (**b**) DC gain

fundamental tones, respectively. The third order intermodulation distortions, IM3 tones, are 24 dB below the fundamentals for 10 mV input signal levels.

The transfer function of the SNT CMOS amplifier including all parasitic resistors is given in Fig. 4.4a. Similarly, the slope of the transfer characteristics representing the DC voltage gain is shown in Fig. 4.4b. The large-signal transient

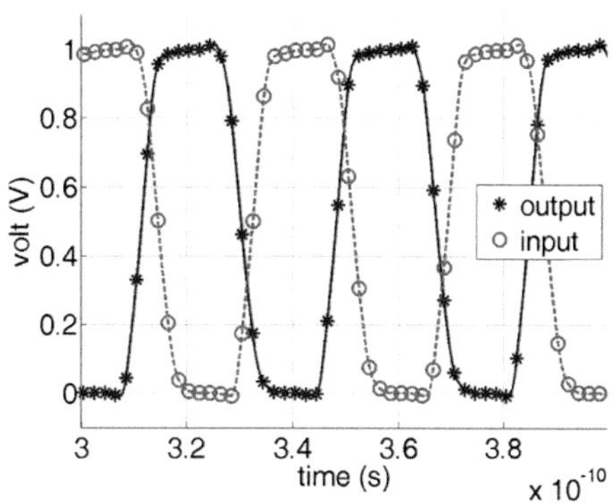

Fig. 4.5 The amplifier large-signal transient response

Table 4.1 Characteristics of SNTs with BSIMSOI modeling

Parameters	NMOS	PMOS
I_{off} ($V_{GS}=0$ V and $V_{DS}=1$ V)	540 pA	80 pA
I_{ds} ($V_{GS}=0.5$ V and $V_{DS}=0.5$ V)	2 μA	0.7 μA
g_m ($V_{GS}=0.5$ V and $V_{DS}=0.5$ V)	14 μA/V	8 μA/V
r_{ds} ($V_{GS}=0.5$ V and $V_{DS}=0.5$ V)	400 kΩ	1.1 MΩ
f_τ	36 THz	25 THz
F_{max}	120 THz	100 THz
CMOS amplifier area	70 × 14 nm	
Amplifier power dissipations	1.64 μW	
3 dB bandwidth	500 GHz	
Third order intermodulation distortions	−24 dBm for $f_2 - f_1 = 10$ GHz	
Small-signal DC gain	−6.5	
Large-signal delay	2.2 ps	
Large-single rise time	5.4 ps	
Large-single fall time	4.7 ps	
Rail-to-swing	1 V	

response of this amplifier acting as an inverter is shown in Fig. 4.5. Seven inverters are cascaded in a closed loop feedback configuration to be able to obtain the input and the output waveforms of a single-stage amplifier in this figure. The initial condition of the circuit ensures self-oscillation with an oscillation frequency of 30 GHz. The gate delay of each SNT inverter is 2.2 ps; the rise and fall times are 5.4 ps and 4.7 ps, respectively. These values are obtained by considering all parasitic values of the layout.

The summary of the SNT amplifier characteristics is listed in Table 4.1.

4.4 Differential SNT Amplifier

4.4.1 A Single-Stage Differential Amplifier Design

Operational amplifiers are one of the most useful building blocks in analog integrated circuits. Since differential amplifiers occupy the input stage of any operational amplifier, their circuit characteristics and performance become crucial in an operational amplifier design. A typical differential amplifier with a current mirror circuit and its layout is shown in Fig. 4.6. The low frequency small-signal model of the amplifier is shown in Fig. 4.7.

In Fig. 4.7, the resistor R_3 is given by

$$R_3 = R_{s3} + \left(r_{ds3} // \frac{1}{g_{m3}} \right) \tag{4.6}$$

For $G_s \gg g_m \gg g_{ds}$, the voltage gain is approximated by

$$\frac{V_{out}}{V_{in}} \approx g_{m2} r_{ds2} \tag{4.7}$$

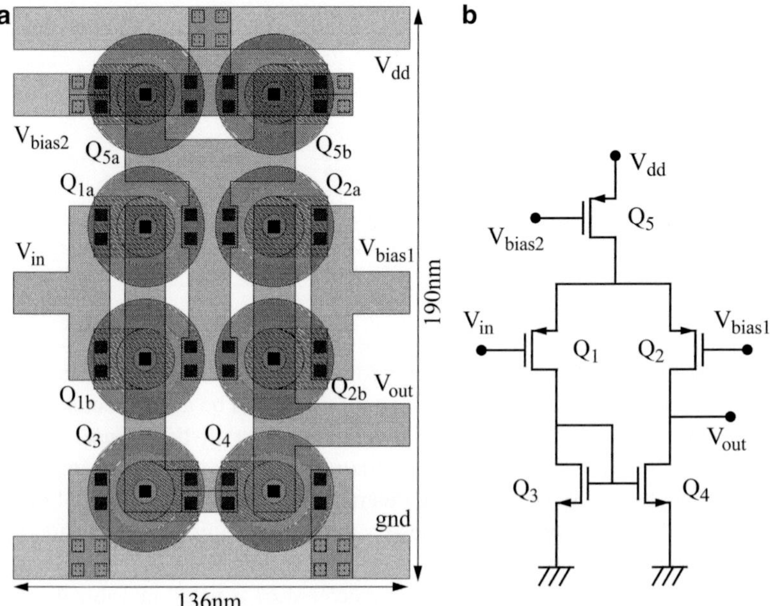

Fig. 4.6 Differential SNT amplifier (**a**) layout and (**b**) schematic

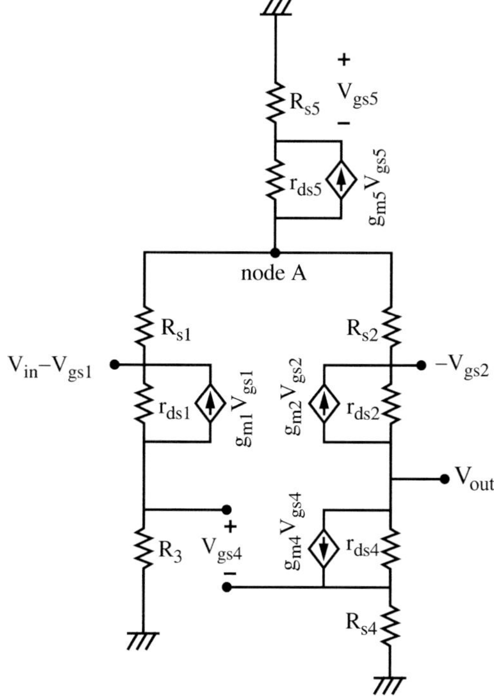

Fig. 4.7 Low frequency small-signal model

For $G_s \gg g_m \gg g_{ds}$, the output resistance is approximately given by

$$R_{out} \approx \left[r_{ds2}\left(1+\frac{g_{m2}}{g_{m1}}\right) \right] // r_{ds4} \tag{4.8}$$

The layout of the differential pair amplifier in Fig. 4.6a is implemented by three metallization layers serving as local interconnects. All interconnect parasitic components are extracted and added to amplifier netlist for post-layout circuit simulations. The PMOS transistors at the inputs of the differential amplifier are implemented by a parallel combination of two PMOS SNTs to ensure a large transconductance. The width of the metal interconnects is selected to be 14 nm to reduce their resistivity and four vias are used to connect metal 2 and metal 3 layers to minimize the ohmic loss. Each 4 nm by 4 nm via offers 400 Ω resistance. A 14 nm by 14 nm overlap capacitance between metal 1 and metal 2 layers is 0.2 aF. The layout area of the differential amplifier occupies an area of 136 nm by 190 nm.

4.4.2 The Characteristics of the Differential Amplifier

The frequency response of the SNT differential pair amplifier is shown in Fig. 4.8. The amplifier provides a gain of 16 with the first pole located at 100 GHz and the second pole located at 100 THz. To attain high accuracy in the transfer functions of

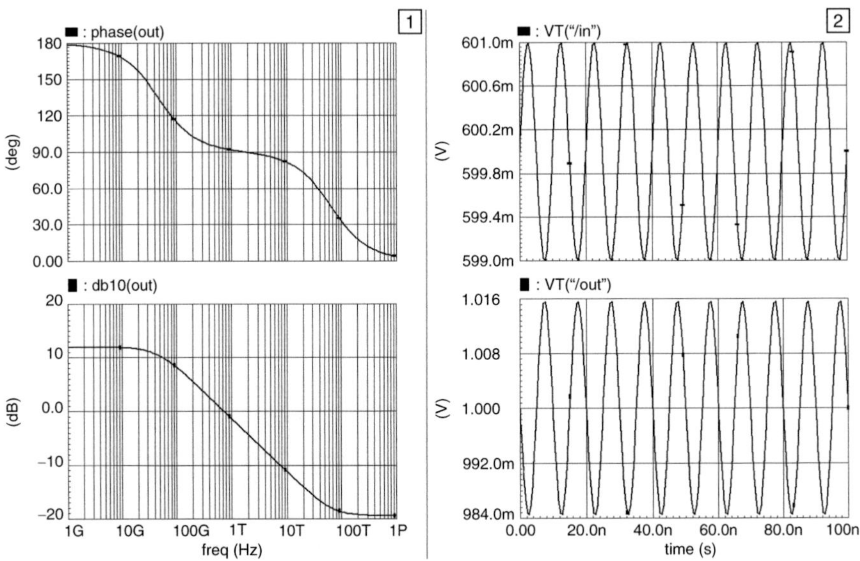

Fig. 4.8 Amplifier small-signal frequency response

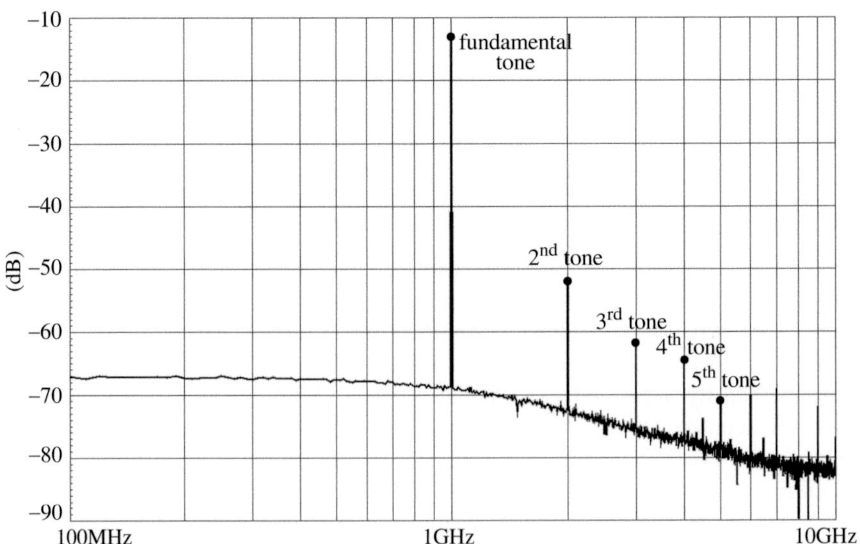

Fig. 4.9 Amplifier output spectrum

various analog circuits such as switch capacitor filters and amplifiers, it may be necessary to cascade multiple such amplifier stages in nested Miller architectures to be able to achieve a voltage gain higher than 1000.

The spectrum of the output waveform of the amplifier is shown in Fig. 4.9. The amplifier produces good linearity with a total harmonic distortion of about 3 % for

Table 4.2 Characteristics of the SNT differential pair amplifier

Parameters	NMOS	PMOS
I_{off} ($V_{GS}=0$ V and $V_{DS}=1$ V)	540 pA	80 pA
I_{ds} ($V_{GS}=0.5$ V and $V_{DS}=0.5$ V)	2 μA	0.7 μA
g_m ($V_{GS}=0.5$ V and $V_{DS}=0.5$ V)	14 μA/V	8 μA/V
R_{ds} ($V_{GS}=0.5$ V and $V_{DS}=0.5$ V)	400 kΩ	1.1 MΩ
f_τ	36 THz	25 THz
F_{max}	120 THz	100 THz
Supply voltage	1.8 V	
Maximum output linear signal swing	0.5 V	
Input DC voltage level	0.6 V	
Voltage gain (at 1 GHz)	16	
Phase margin	>90°	
Unity voltage gain cutoff frequency	5.1 THz	
Third order intermodulation distortion (10 mV two-tone signals with 1 GHz spacing)	−24 dBm	
Second order harmonic distortion	−40 dBm	
Third order harmonic distortion	−52 dBm	
Total harmonic distortion	3 %	
Load capacitor	20 aF	
Power dissipation	5 μW	
Area	136 × 190 nm	

±233 mV output swing. Such a high linearity is due to the source resistance, R_s, acting as the degeneration resistance and thereby minimizing the harmonic distortion of the differential PMOS SNTs at the input.

The summary of the single-stage SNT amplifier is listed in Table 4.2.

4.5 Multi-stage SNT Operational Amplifier

4.5.1 A Two-Stage Operational Amplifier Design

Intrinsic BSIMSOI transistor models are obtained after matching the input and output I–V characteristics of NMOS and PMOS SNTs with the I–V characteristics obtained from the three-dimensional device simulations as mentioned earlier. Parasitic RC components are subsequently added to the intrinsic models to create realistic extrinsic SPICE models used for circuit simulations. Chapter 3 shows the dominant RC parasitics of an SNT superimposed on its three-dimensional structure. In this figure, the parasitic capacitance between the source contact and the metal gate is denoted as C_{gs2}. The parasitic capacitance between the metal gate and the concentric source contact, C_{gs1}, is considered the largest dominant capacitor for the SNT. The gate-drain capacitors, C_{gd1} and C_{gd2}, and the drain-source capacitors,

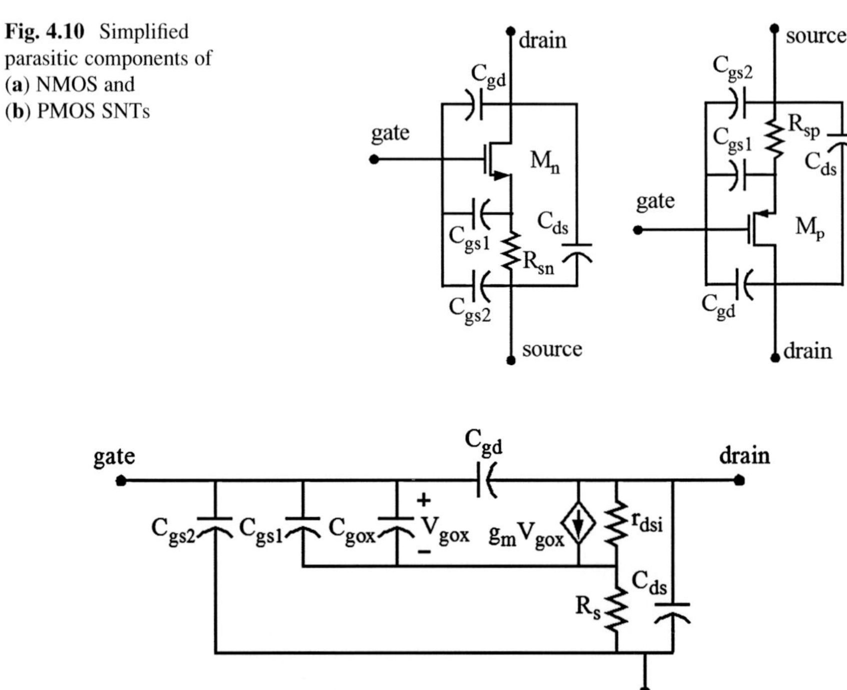

Fig. 4.10 Simplified parasitic components of (**a**) NMOS and (**b**) PMOS SNTs

Fig. 4.11 Linearized small-signal model of SNT

C_{ds1} and C_{ds2}, can be lumped together to form C_{gd} and C_{ds}, respectively. Compared to planar bulk transistors, C_{gd} produces a very small value. In addition, there is no junction to bulk capacitance; therefore, C_{ds} becomes quite linear [1–3]. The well resistance, R_s, can be large and is a major drawback in vertical SNTs compared to planar transistors. The magnitude of this resistance can be reduced drastically by placing a concentric (ring shape) source contact in parallel with the well as discussed in Chapter 3. If the transistor is properly designed to ensure $G_s \gg g_m \gg g_{ds}$ (g_m is the intrinsic transconductance and g_{ds} is the intrinsic output conductance), the performance of the vertical SNTs surpasses those of the planar transistors for analog circuits.

The simplified parasitic RC components of the NMOS and PMOS SNTs are shown in Fig. 4.10 with lumped capacitors of C_{gd} and C_{ds}.

For simplified hand calculations and the amplifier AC parameters, the small-signal model in Fig. 4.11 can be used. This model is also helpful to compute the DC voltage gain and the output resistance of various amplifier stages at low frequencies.

Operational amplifiers offer differential amplifiers at the input stage followed by common-source amplifiers, cascode amplifiers and common-drain amplifiers in the later stages. The operation and the characteristics of an SNT differential amplifier

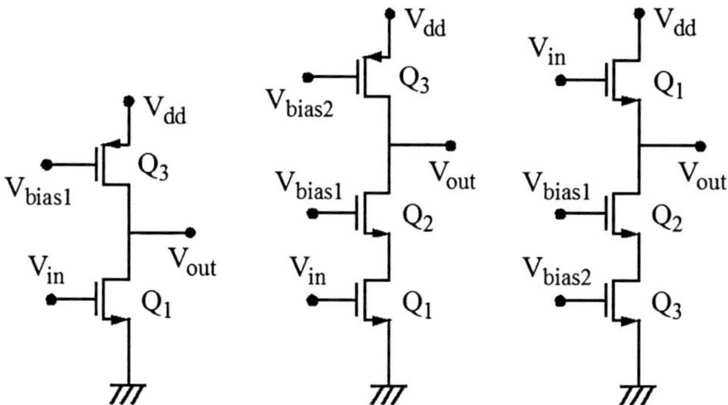

Fig. 4.12 Common-source, cascode, and common-drain amplifiers

were studied in the previous section. According to this study, the differential amplifier produced a gain of 16 at the first pole located at 100 GHz. However, to achieve high accuracy in the transfer functions of switch capacitor filters and amplifiers, it is important that operational amplifiers provide a voltage gain higher than 1000. Therefore, using only one differential pair at the input stage is not sufficient to implement a high-gain voltage amplifier; additional amplifying stages are most likely required. The second amplifying stage can also be used to improve the phase margin of the operational amplifier and ensure stability when using the amplifier in a closed loop configuration.

Common-source amplifiers, cascode amplifiers or common-drain amplifiers are the types of the amplifiers that follow the differential amplifier as shown in Fig. 4.12. Before designing a multi-stage SNT operational amplifier, the characteristics and the performance figures of each type need to be investigated.

Common-source amplifiers are used in the second stage of an operational amplifier and function as the Miller compensating stage. The low frequency small-signal model of the common-source amplifier is shown in Fig. 4.13.

The Kirchoff's current law at output branch results in

$$G_{s2}V_{gox2} = G_{s1}\left(V_{in} - V_{gox1}\right) \tag{4.9}$$

The output voltage V_{out} can be expressed by

$$V_{out} = (1 + r_{ds1}G_{s1})V_{in} - [1 + r_{ds1}(G_{s1} + g_{m1})]V_{gox1} \tag{4.10}$$

and

$$V_{out} = -[1 + r_{ds2}(G_{s2} + g_{m2})]V_{gox2} \tag{4.11}$$

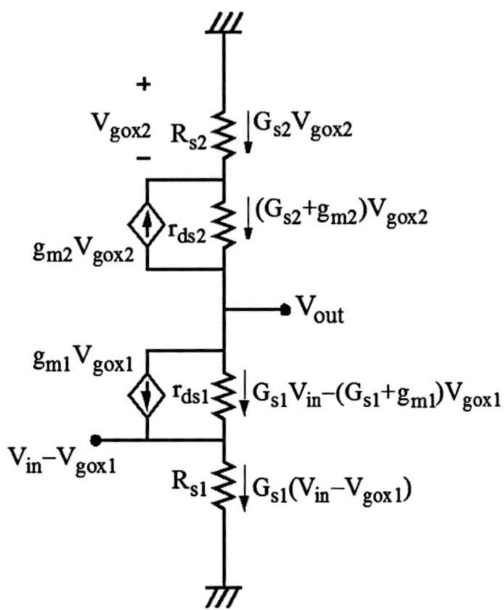

Fig. 4.13 Low frequency small-signal model of the common-source amplifier

Then the voltage gain of the common-source amplifier is given by

$$\frac{V_{out}}{V_{in}} = \frac{-g_{m1}r_{ds1}\left[1 + r_{ds2}(G_{s2} + g_{m2})\right]}{\left[1 + r_{ds2}(G_{s2} + g_{m2})\right] + \frac{G_{s2}}{G_{s1}}\left[1 + r_{ds1}(G_{s1} + g_{m1})\right]} \tag{4.12}$$

For $G_s \gg g_m \gg g_{ds}$, the voltage gain is approximately given by

$$\frac{V_{out}}{V_{in}} = -g_{m1}(r_{ds1}//r_{ds2}) \tag{4.13}$$

The intrinsic gain, $g_m.r_{ds}$, can be as high as 15 for a common-source amplifier and it is not enough to use it for the Miller stage. SNTs have very high unity-gain cutoff frequency and very low current handling capability. Therefore, using a large Miller capacitance is not feasible to achieve frequency compensation. Therefore, other amplifier structures which can provide much larger gain need to be investigated to implement the Miller stage.

A cascode amplifier, as the second alternative, inverts the input signal as a common-source amplifier but with a much larger gain. The low frequency small-signal model of the cascode amplifier is shown in Fig. 4.14.

The Kirchoff's current law at output branch results in

$$G_{s3}V_{gox3} = G_{s1}\left(V_{in} - V_{gox1}\right) \tag{4.14}$$

The output voltage V_{out} can be expressed by

$$V_{out} = -\left[1 + r_{ds3}(G_{s3} + g_{m3})\right]V_{gox3} \tag{4.15}$$

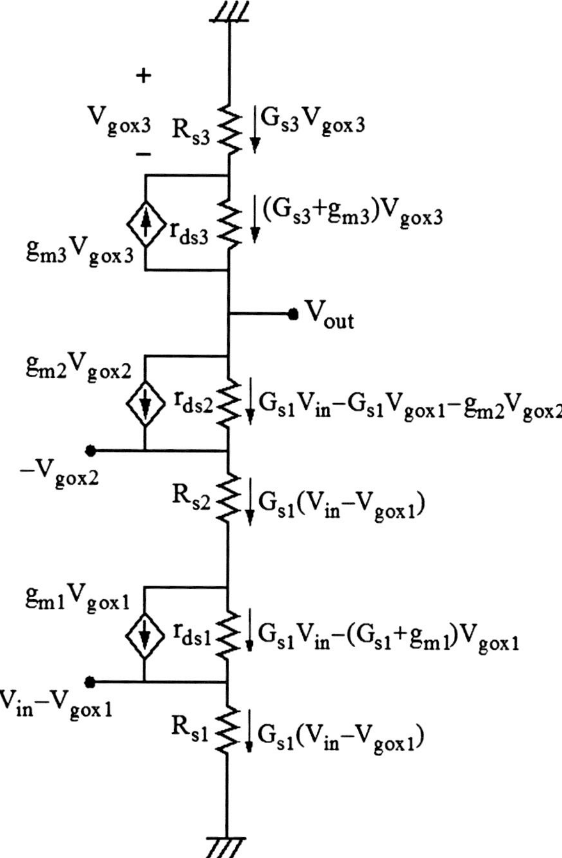

Fig. 4.14 Low frequency small-signal model of the cascode amplifier

and

$$V_{out} = r_{ds2}G_{s1}V_{in} - r_{ds2}G_{s1}V_{gox1} - (1 + r_{ds2}g_{m2})V_{gox2} \tag{4.16}$$

where

$$V_{gox2} = -[1 + G_{s1}(r_{ds1} + R_{s2})]V_{in} + [1 + G_{s1}(r_{ds1} + R_{s2}) + g_{m1}r_{ds1}]V_{gox1} \tag{4.17}$$

The voltage gain of the cascode amplifier is given by

$$\frac{V_{out}}{V_{in}} = \frac{-G_{s1}[1 + r_{ds3}(G_{s3} + g_{m3})][(1 + g_{m2}r_{ds2})(2R_{s1} + 2R_{s2} + r_{ds1}(2 + g_{m1}R_{s1}))]}{G_{s3}[(1 + g_{m2}r_{ds2})(R_{s1} + R_{s2} + r_{ds1}(1 + g_{m1}R_{s1})) - r_{ds2}] - [1 + r_{ds3}(G_{s3} + g_{m3})]} \tag{4.18}$$

For $G_s \gg g_m \gg g_{ds}$, the voltage gain is approximated as

$$\frac{V_{out}}{V_{in}} = \frac{-2g_{m2}r_{ds1}r_{ds2}r_{ds3}G_{s1}}{g_{m1}g_{m2}r_{ds1}r_{ds2}R_{s1} - r_{ds3}} \tag{4.19}$$

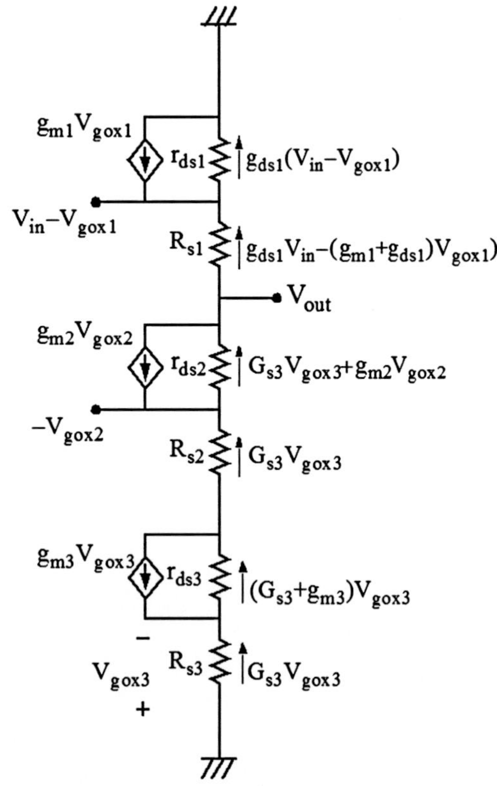

Fig. 4.15 Low frequency small-signal model of the buffer amplifier

The voltage gain can be very high due to a small value in the denominator.

The output resistance of the cascode amplifier is given by

$$R_{out} = [R_{s2} + g_{m2}r_{ds1}r_{ds2}(1 + g_{m1}R_{s1})]//[r_{ds3}(1 + g_{m3}R_{s3})] \tag{4.20}$$

For $G_s \gg g_m \gg g_{ds}$, the output resistance can be approximated as

$$R_{out} \approx r_{ds3}(1 + g_{m3}R_{s3}) \tag{4.21}$$

The SNT operational amplifier may have to drive high capacitive loads that belong to other analog stages. Therefore, it is very important to isolate the Miller stage from the load by using a buffer stage. A common-drain amplifier can be used for this purpose. The low frequency small-signal model of a common-drain buffer amplifier is shown in Fig. 4.15.

The Kirchoff's current law at output branch results in

$$G_{s3}V_{gox3} = -g_{ds1}V_{in} + (g_{ds1} + g_{m1})V_{gox1} \tag{4.22}$$

The output voltage V_{out} can be expressed by

$$V_{out} = (1 + R_{s1}g_{ds1})V_{in} - [1 + R_{s1}(g_{m1} + g_{ds1})]V_{gox1} \tag{4.23}$$

and

$$V_{out} = -(1 + r_{ds2}g_{m2})V_{gox2} - G_{s3}r_{ds2}V_{gox3} \tag{4.24}$$

where

$$V_{gox2} = -[1 + G_{s3}(r_{ds3} + R_{s2}) + g_{m3}r_{ds3}]V_{gox3} \tag{4.25}$$

The voltage gain of the amplifier is given by

$$\frac{V_{out}}{V_{in}} = \frac{g_{m1}[-r_{ds2} + (1 + g_{m2}r_{ds2})(r_{ds3} + R_{s2} + R_{s3} + g_{m3}r_{ds3}R_{s3})]}{1 + (g_{m1} + g_{ds1})[R_{s1} - r_{ds2} + (1 + g_{m2}r_{ds2})(r_{ds3} + R_{s2} + R_{s3} + g_{m3}r_{ds3}R_{s3})]} \tag{4.26}$$

For $G_s \gg g_m \gg g_{ds}$, the voltage gain is approximated as

$$\frac{V_{out}}{V_{in}} \approx \frac{g_{m1}}{g_{m1} + g_{ds1}} \tag{4.27}$$

and this value is very close to unity.

The output resistance of the amplifier is given by

$$R_{out} = \left[R_{s1} + \left(r_{ds1} // \frac{1}{g_{m1}}\right)\right] // (g_{m2}r_{ds2}r_{ds3}) \tag{4.28}$$

For $G_s \gg g_m \gg g_{ds}$, the output resistance is approximated as

$$R_{out} \approx \frac{1}{g_{m1}} \tag{4.29}$$

and this value becomes very small.

Therefore, a high performance SNT operational amplifier can be implemented by a differential amplifier at the input stage, a cascode amplifier in the second stage and a common-drain amplifier in the final buffer stage as shown in Fig. 4.16.

The biasing control resistance, R_0, is adjusted to create a reference current of 670 μA for I_{ds13}, I_{ds23}, I_{ds43}, and I_{ds51}, and 1.4 mA for I_{ds33} in the current mirror circuit.

The overall low frequency voltage gain of the operational amplifier is given by

$$\frac{V_{out}}{V_{in}} \approx g_{m34}r_{ds34}\frac{-2g_{m42}r_{ds41}r_{ds42}r_{ds43}G_{s41}}{(g_{m41}g_{m42}r_{ds41}r_{ds42}R_{s41} - r_{ds43})}\frac{g_{m53}}{(g_{m53} + g_{ds53})} \tag{4.30}$$

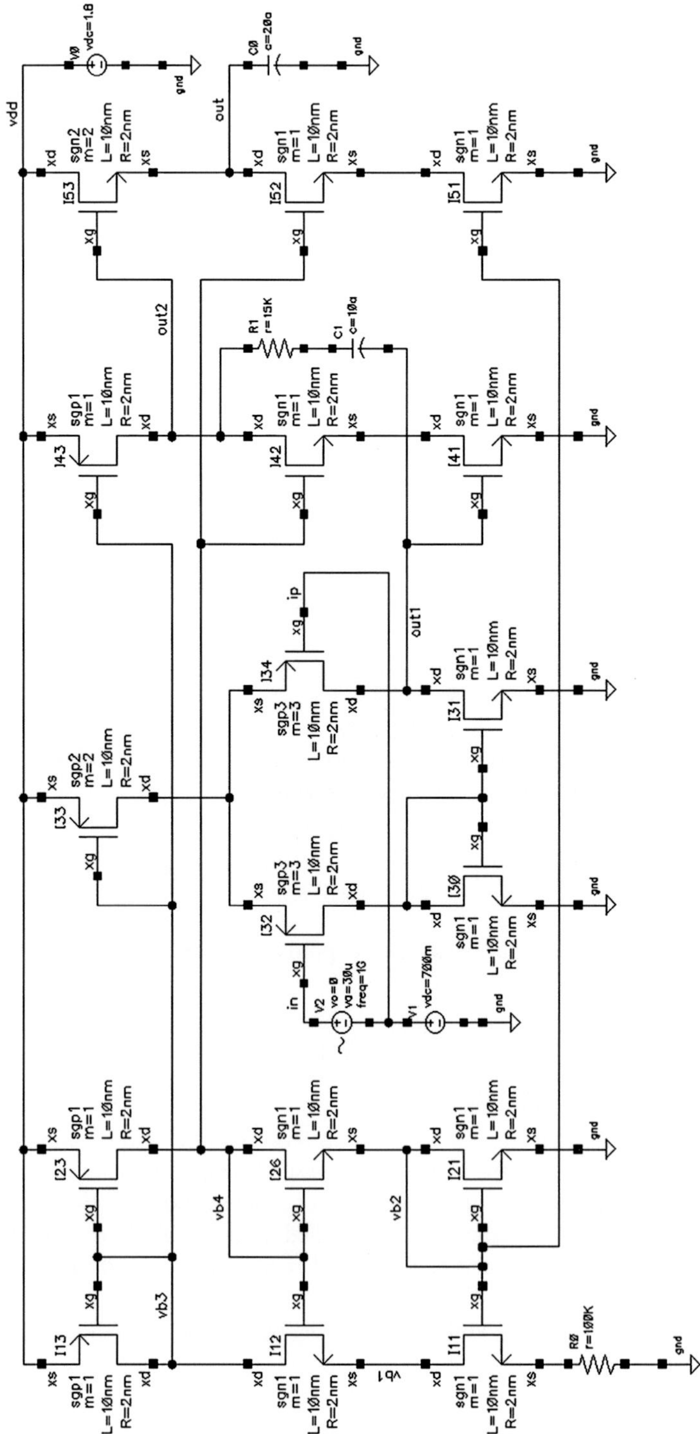

Fig. 4.16 The schematic of the two-stage operational amplifier

The location of the first pole is given by

$$f_L \approx \frac{1}{2\pi C_1 \frac{2g_{m42}r_{ds41}r_{ds42}r_{ds43}G_{s41}}{(g_{m41}g_{m42}r_{ds41}r_{ds42}R_{s41}-r_{ds43})}\left[r_{ds34}\left(1+\frac{g_{m34}}{g_{m32}}\right)//r_{ds31}\right]} \tag{4.31}$$

and the location of the second pole determined by the load capacitor is given by

$$f_H \approx \frac{g_{m53}}{2\pi C_0} \tag{4.32}$$

The slew rate (SR) is computed as

$$SR \approx \frac{I_{ds33}}{C_1} \tag{4.33}$$

4.5.2 Characteristics of the Operational Amplifier

The impact of the Miller capacitance on the location of the poles and its effect on the frequency response of the compensated and uncompensated operational amplifiers are shown in Fig. 4.17. The lead compensation produced by a series combination of a resistor, R_1, and a capacitor, C_1, helps to improve the unity voltage gain cutoff frequency of the operational amplifier without dissipating any power.

The layout of the SNT operational amplifier is shown in Fig. 4.18. The input transistors in the differential amplifier are implemented by a parallel combination of three PMOS SNTs to produce large transconductance and to be able to construct the Miller stage with NMOS SNTs. The biasing control resistor, R_0, is realized by an 8 nm wide and 320 nm long N-well. The frequency compensation is achieved by using a resistor, R_1, and a capacitor, C_1. In the compensation circuit, C_1 is implemented by 150 nm wide and 63 nm long metal 1 and metal 2 interconnect sandwich with 36 nm spacing. The resistor R_1 is implemented by 8 nm wide and 48 nm long N-well. The width of the metal interconnects is selected to be 14 nm to

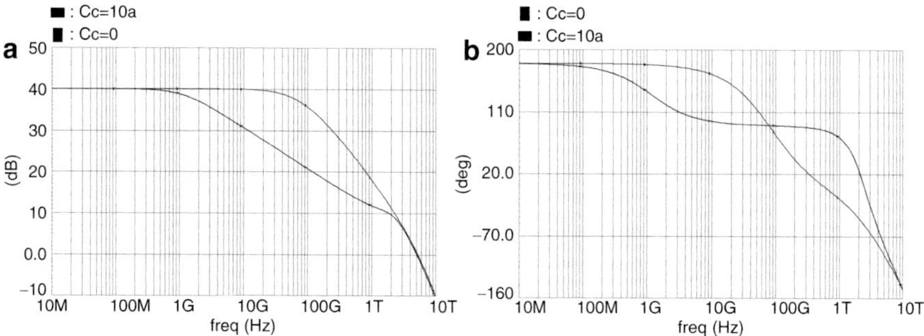

Fig. 4.17 Compensated and uncompensated (**a**) gain magnitude and (**b**) phase response of the SNT opamp

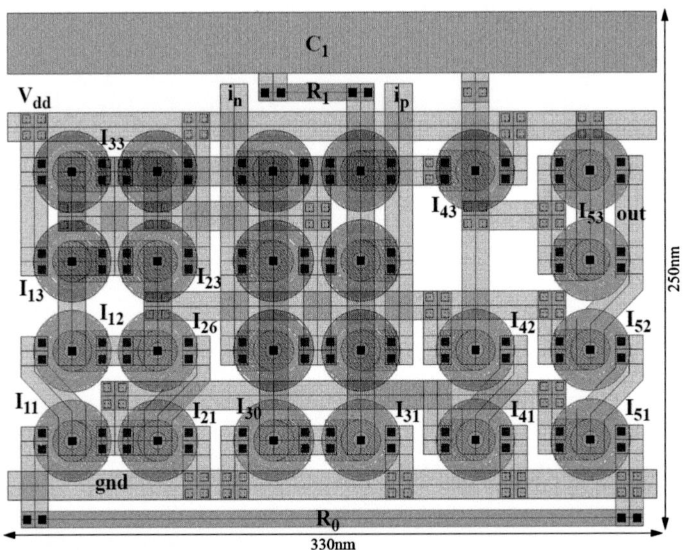

Fig. 4.18 The layout of the SNT operational amplifier

reduce the overall resistance, and four parallel vias are used to connect metal 1 layer to metal 2 layer to minimize the ohmic loss. Each 4 nm by 4 nm via produces 400 Ω. Each 14 nm by 14 nm overlap capacitance formed between metal 1 and metal 2 layers is 0.2 aF. The layout area of the operational amplifier including the biasing circuit and all compensation components occupies an area of 330 nm by 250 nm.

The post-layout frequency response of the SNT operational amplifier is shown in Fig. 4.19. This amplifier reveals a high unity voltage gain cutoff frequency at 5.1 THz but only dissipates 7.2 μW. The operational amplifier has a phase margin better than 90° for frequencies less than 1 THz and achieves a very stable operation.

The open loop transient response of the operational amplifier is shown in Fig. 4.20. The low frequency voltage gain of the amplifier is approximately 7760. For an input swing with 30 μV peak amplitude, the output of the amplifier produces 233 mV. The operational amplifier has good linearity characteristics and exhibits a total harmonic distortion of 3 % for ±233 mV output swing. This high linearity can be attributed to the source resistance, R_s, acting as the degeneration resistance, minimizing the harmonic distortions of each amplifier stage.

The Common-Mode Rejection Ratio (CMRR) and the Power Supply Rejection Ratio (PSRR) of the operational amplifier are shown in Fig. 4.21. The CMRR has a corner frequency at 100 GHz and the PSRR has a corner frequency at 1 GHz. The CMRR achieves 40 dB and the PSRR achieves 54 dB signal rejection.

The post-layout characteristics of the operational amplifier are listed in Table 4.3.

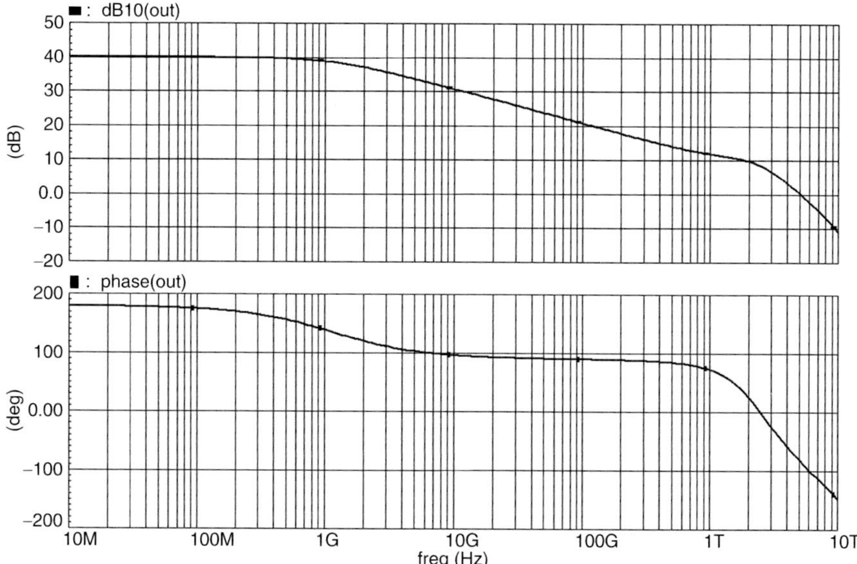

Fig. 4.19 Frequency response of the SNT operational amplifier

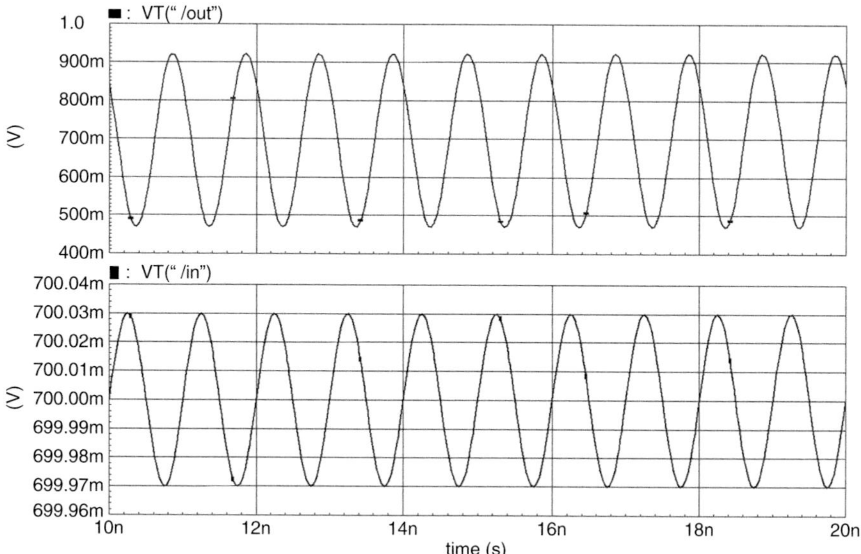

Fig. 4.20 Transient input and output waveforms of the operational amplifier

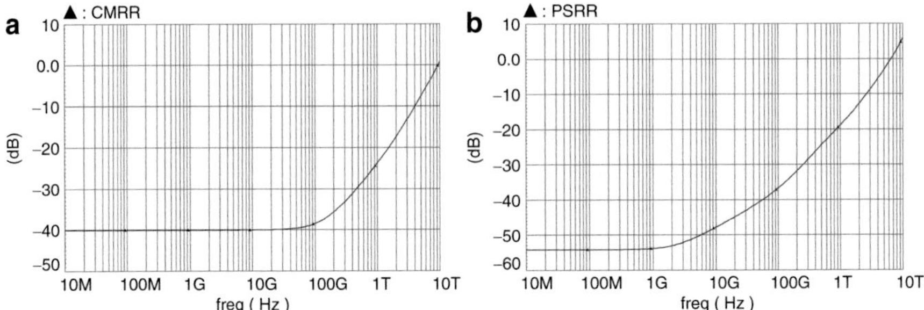

Fig. 4.21 (**a**) CMRR and (**b**) PSRR of the operational amplifier

Table 4.3 Post-layout characteristics of the two-stage SNT operational amplifier

Operational amplifier characteristics	
Supply voltage	1.8 V
Biasing circuit power dissipation	2.4 μW
Total power dissipation	4.8 μW
Maximum output linear signal swing	0.5 V
Input DC voltage level	0.7 V
Open loop voltage gain (at 1 GHz)	7760 or 38.9 dB
Open loop phase margin (at 1 GHz)	42°
Open loop unity voltage gain cutoff frequency	5.1 THz
Load capacitor	20 aF
Common-mode rejection ratio	40 dB
Power supply rejection ratio	54 dB
Total harmonic distortion	3 %
Slew rate (high to low)	2 V/ns
Slew rate (low to high)	2 V/ns
Area	0.0825 μm^2

4.6 Summary

The single-stage CMOS amplifier is a simple but a good circuit platform to show the capabilities of SNTs in large-signal and small-signal AC domains. In large-signal domain when the amplifier is used as an inverter, the circuit reveals 2.2 ps delay, 5.4 ps rise time and 4.7 ps fall time while oscillating at 30 GHz. In small-signal domain, the amplifier acts as a common-source amplifier if a proper DC level is introduced and a small AC source is applied to the input. When used as a single-stage common-source amplifier, the circuit dissipates 1.64 μW total power and produces a 500 GHz bandwidth with a −6.5 gain. The third order intermodulation distortion tones for a two-tone input signal become −24 dBm with 10 mV amplitude and 10 GHz frequency spacing.

Differential pair amplifiers provide unique circuit characteristics compared to all the other amplifier structures. Because there are two different AC signal paths from input to output, the gain of an amplifier in differential mode is much higher compared to the gain of an amplifier in common mode. This type also eliminates the amplification of noise and cross-talk because of the differential inputs. Besides high gain and low power, differential SNT amplifiers can achieve a very high common-mode rejection ratio which makes them good candidates for implementing operational amplifiers. The differential amplifier studied in this chapter dissipates 5 μW power and produces a 5 THz bandwidth with a voltage gain of 16. It produces a linear output voltage swing of 0.5 V and a total harmonic distortion better than 3 % from a 1.8 V power supply and a 20 aF capacitive load. The second and third order harmonic distortions of the amplifier are −40 and −52 dBm, respectively, and the third order intermodulation is −24 dBm for a two-tone input signal with 10 mV amplitude and 10 GHz frequency spacing.

The SNT operational amplifier consists of a differential amplifier at the input stage followed by a cascode amplifier. A common-drain amplifier can be used as a third stage in an operational amplifier to isolate the Miller stage from the load. This buffer stage also decreases the output impedance to be able to drive low resistive loads. The amplifier is frequency compensated for oscillation-free, stable operation with a 1.8 V power supply. It has a voltage gain of 40 dB and a phase margin of 42°. The current gain cutoff frequency is 5.1 THz and the amplifier produces 40 dB common-mode rejection ratio and 54 dB power supply rejection ratio with a slew rate of 2 V/ns. The layout area of a typical SNT operational amplifier is 320 nm by 250 nm.

References

1. Bindal A, Naresh A, Yuan P, Nguyen KK, Hamedi-Hagh S (2007) The design of dual work function CMOS transistors and circuits using silicon nanowire technology. IEEE J Nanotechnol 6:291–302
2. Bindal A, Hamedi-Hagh S (2006) The impact of silicon nanowire technology on the design of single work function CMOS transistors and circuits. Inst Phys J Nanotechnol 17:4340–4351
3. Hamedi-Hagh S, Bindal A (2008) Characterization of nanowire CMOS amplifiers using fully depleted surrounding gate transistors. J Nanoelectron Optoelectron 3(3):281–288

Chapter 5
Radio Frequency (RF) Applications

5.1 Introduction

This chapter presents the application of silicon nanowire technology on another analog domain: RF receivers. RF receivers contain two important functional blocks. The first block is the down-converter unit. This block consists of a low noise amplifier, a mixer, and an oscillator. It is mainly used to translate a high frequency signal to a lower frequency signal for baseband processing. The second block is the Variable Gain Amplifier (VGA). Because the signal strength largely depends on the distance between the receiver and transmitter, the VGA regulates the signal strength before the re-conditioned signal arrives at the input of the baseband processor. The design approach and and the circuits incorporating SNTs are fully examined in this chapter to design and analyze both of these units. The layout of each SNT includes a re-engineered source structure as well as extra gate and source contacts to minimize high terminal resistances as they deteriorate the overall noise factor of the RF receiver and make the input impedance matching difficult.

5.2 Brief Description of Transistor Design and Modeling

The 2 nm radius and 10 nm channel length transistor body dimensions are the result of minimum intrinsic transient times and minimum static and dynamic power dissipations for each NMOS and PMOS SNT as discussed in Chapter 3. The BSIMSOI device models in Chapter 3 are also based on these particular device dimensions and they are used for all the circuit simulations in this chapter.

A. Bindal, S. Hamedi-Hagh, *Silicon Nanowire Transistors*,
DOI 10.1007/978-3-319-27177-4_5

5.3 RF Receiver Front End

5.3.1 Receiver Topology

In wireless communications, designing an optimal Radio Frequency (RF) receiver that consumes minimal power without sacrificing signal quality is a challenging task [1–3]. An RF receiver consists of three parts: the Low Noise Amplifier (LNA), Mixer, and Voltage-Controlled Oscillator (VCO) as shown in Fig. 5.1. The signal received at the antenna is initially amplified by the LNA. The amplified signal is transformed to a lower frequency signal by an operation called down-conversion. During this transformation, the Local Oscillator unit (LO) allows the channel selection of RF signals when used with a mixer. Following the down-conversion operation, the interference components of the input signal are eliminated by a band-pass filter before the Intermediate Frequency (IF) signal is sent to data converters and the baseband processor for data extraction.

5.3.2 LC Tank Voltage-Controlled Oscillator (VCO)

A Voltage-Controlled Oscillator (VCO) is usually designed by an LC tank to achieve low-phase noise in wireless systems. A resonator usually implemented by a parallel combination of an inductor and a capacitor (LC tank). The VCO with LC tank can be considered as two single-port networks combined together. The first single-port circuit represents the resonator that determines the frequency of oscillation. The second single-port circuit represents the active circuit that cancels the losses in the LC tank as shown in Fig. 5.2.

To achieve steady-state oscillations, the following conditions must be met:

1. Magnitude of the loop gain (α) must be equal to 1; the phase shift in the loop must be equal to zero, yielding no imaginary component.

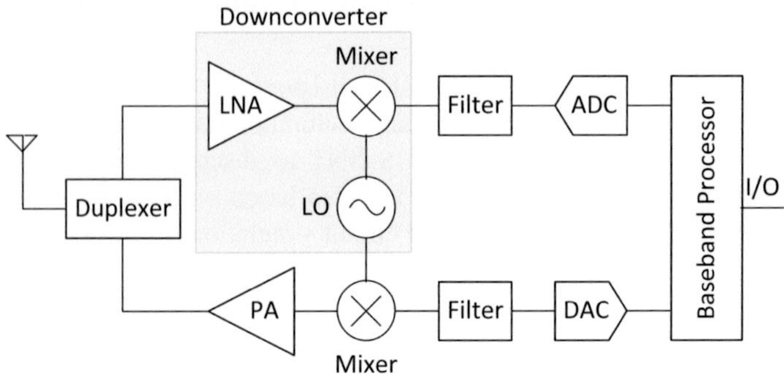

Fig. 5.1 RF receiver down-converter block

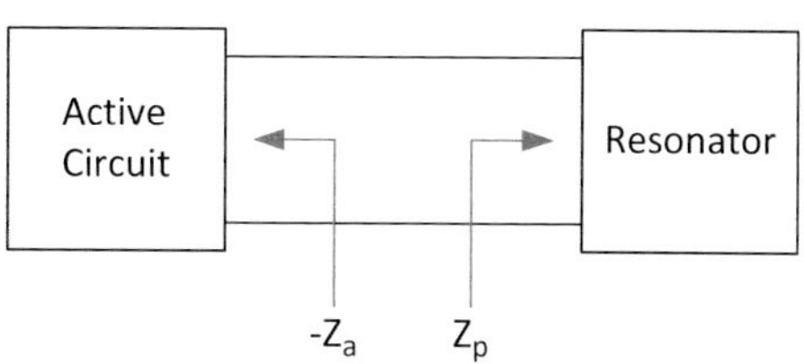

Fig. 5.2 One-port view of an oscillator

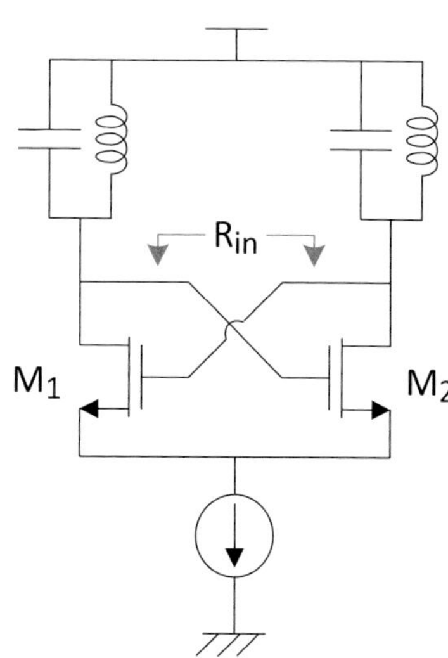

Fig. 5.3 A cross-coupled differential topology

2. The equivalent parallel impedance, Z_p, of the resonator comprising two LC tank circuits must be balanced with the negative impedance, $-Z_a$, produced by the active circuit.

When these two conditions are satisfied, the circuit becomes lossless and generates oscillations. The oscillator can use two LC resonators to operate differentially as shown in Fig. 5.3. This configuration is called a cross-coupled differential oscillator or a negative-g_m oscillator and it can be considered a single-port implementation in this figure. Negative resistance seen at the drain of M1 and M2 is expressed by

$$R_{in} = -\frac{2}{g_m} = Z_a \tag{5.1}$$

where g_m is transconductance of each transistor.

If R_{in} is less than or equal to the equivalent parallel impedance Z_p of the tank, the circuit starts to oscillate.

In reality, there are losses associated with capacitors, inductors, and MOS transistors. In integrated VCOs, the inductors have low quality factor that dominates the losses of the VCO tank. The quality factor, Q_L, of the inductor is given by

$$Q_L = \frac{\omega_o L}{R} \tag{5.2}$$

where ω_o is the oscillation frequency, L is the value of the inductance and R is the inductor's equivalent series resistance. For a conventional LC tank circuit, the oscillation frequency of the circuit is given as

$$\omega_0 = \frac{1}{\sqrt{LC}}\sqrt{1 - \frac{R^2C}{L}} \tag{5.3}$$

This is under the assumption that all the capacitors and MOS transistors in the circuit are ideal. Under the same assumption, g_m of each MOS transistor must satisfy

$$g_m \geq \frac{RC}{L} \tag{5.4}$$

For two LC tank circuits in series, the equivalent impedance, Z_p, becomes

$$Z_p = R_p + jX_p = 2(R_s + jX_s) \tag{5.5}$$

where R_p is the equivalent resistance and X_p is the equivalent reactance.

Assuming the transistors operate in saturation, the saturation current, I_{dsat}, and g_m become

$$I_{dsat} = \frac{K_p}{2}\frac{W}{L_{channel}}\left(V_{gs} - V_t\right)^2 = \frac{g_m}{2}\left(V_{gs} - V_t\right) \tag{5.6}$$

and

$$g_m = \frac{dI_{dsat}}{dV_{gs}} = K_p\frac{W}{L_{channel}}\left(V_{gs} - V_t\right) \tag{5.7}$$

where $L_{channel}$ is the minimum channel length, V_t is the threshold voltage, V_{gs} is the gate-to-source voltage, and K_p is a device constant. The width W of each NMOS transistor can thus be determined once g_m is found.

5.3.3 *Mixer*

Mixer is another essential block in the RF receiver. Mixers produce frequency translation by multiplying two signals and their harmonics. Down-conversion mixers implemented in the receiver's path have two different inputs: the RF port

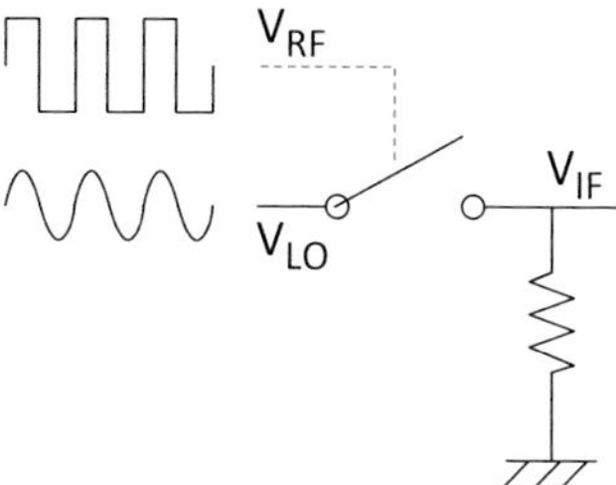

Fig. 5.4 Simple switch used as mixer

and the LO port. The RF port senses the amplified signal coming from the output of the down-converted LNA. The LO port senses the periodic waveform generated by the LC tank in the VCO. This is illustrated in Fig. 5.4 where the output is equal to the RF input when the switch is turned on and equal to zero when the switch is turned off. This operation can also be considered as the multiplication of the RF signal by a rectangular waveform. For most down-conversion mixers, the circuit is linear and time variant with respect to the RF port, and nonlinear and time variant with respect to the LO port. The RF port must have low noise and high linearity to avoid strong intermodulation products caused by the LNA interference amplification.

5.3.4 Low Noise Amplifier (LNA)

Noise is present in any electrical system and caused by numerous factors. At high frequencies, the random movement of electrons in the MOS transistor channel is the main source of noise known as the thermal noise. At lower frequencies, electron traps at the silicon-dioxide channel interface are the main source of noise called the flicker noise.

The received signal through the RF antenna is often weak and needs to be amplified. The signal also contains some noise component which is amplified along with the actual signal. For n number of cascaded two-port devices with noise figures of F1, F2, F3..., Fn and gains of G1, G2, G3..., Gn, the total noise factor of the entire network is given as follows:

$$F_{tot} = F_1 + \frac{F_2}{G_1} + \frac{F_3}{G_1 G_2} + \ldots + \frac{F_n}{G_1 G_2 \ldots G_{n-1}} \tag{5.8}$$

In this equation, the noise content of the first block is directly added to the total noise figure. However, the contribution of each remaining block to the overall noise decreases with the product of the previous amplifier stage gains. However small or large the contribution of each cascaded stage may be, each amplifier contributes to the total noise. Therefore, a special kind of amplifier needs to be designed in order to achieve a low overall noise figure instead of cascaded amplifier stages. The low

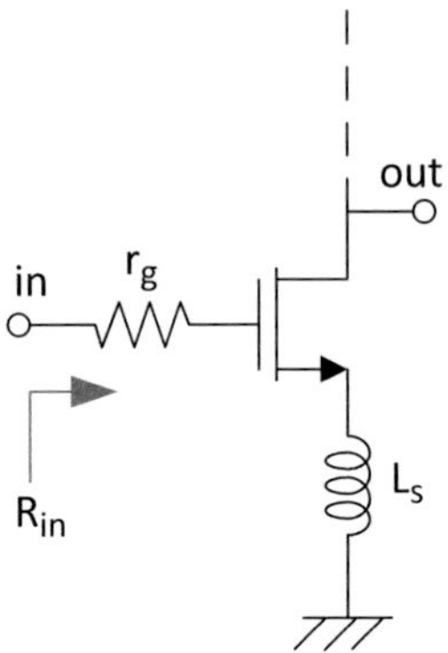

Fig. 5.5 Initial single-stage LNA

noise amplifier is usually connected directly (or through a transmission line) to the antenna. For maximum power transfer between the antenna and the amplifier, impedance matching should be performed. This leads to equating the output impedance of a transmission line to the input impedance of the LNA. The LNA input impedance, Rin, can be controlled by biasing and sizing the transistor as shown in Fig. 5.5.

R_{in} is given as

$$R_{in} = r_g + \frac{L_s}{C_{gs}} g_m \tag{5.9}$$

Here, r_g is the gate resistance, L_s is the inductance connected to the source of the SNT, C_{gs} is the gate-to-source capacitance and g_m is the transconductance of the SNT in Fig. 5.5.

An important aspect of the LNA design is how to optimize the circuit for minimum noise operation and maximum power transfer. When R_{in} is impedance-matched to a 50 Ω transmission line for minimum reflection coefficient and maximum power transfer, the LNA noise factor does not necessarily become minimum. Similarly, a minimum noise factor in LNA does not necessarily guarantee a perfect impedance match, resulting in maximum amount of power delivery to the LNA. Therefore, in order to achieve maximum power transfer with minimal amplifier noise, circuit biasing conditions and transistor sizes are often varied until both impedance mismatch and noise factor are in acceptable levels.

5.3.5 LNA-Mixer-VCO (LMV) Cell

The VCO consists of a pair of LC tank oscillators and a switching pair of NMOS SNT transistors. Cascading the mixer with the LNA input stage is a conventional implementation, while stacking the VCO and the mixer is unconventional and less frequent. Both the VCO and the mixer have a differential transistor pair that

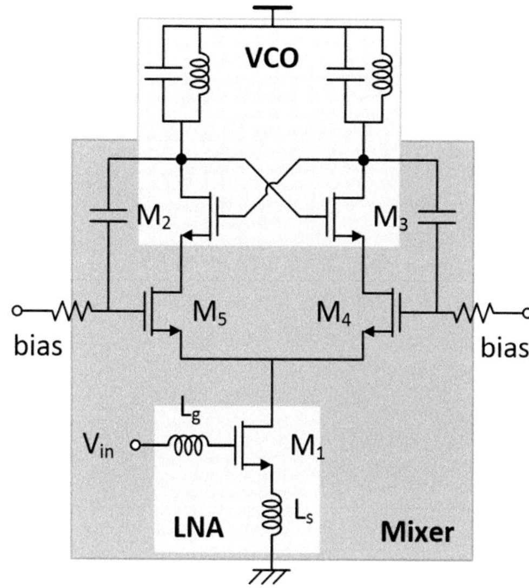

Fig. 5.6 LNA-Mixer-VCO (LMV) cell

behaves like a switch. Therefore, it is advantageous to use a new approach and merge the two blocks into a single block as shown in Fig. 5.6. This circuit is called Self-Oscillating Mixer (SOM).

5.3.6 LC Tank Oscillator as a Mixer

The conventional LC tank oscillator shown in Fig. 5.7 can also act as a mixer since any RF signal in the VCO bias current is naturally down-converted to an IF signal by the switching pair, M2-M3. In the down-converter topology, the high-Q LC tank attenuates the low frequency components of V_{out}, thus preventing any IF amplification. The same mechanism is used for up-converting the DC current of M1 to the LO frequency where the LC tank oscillator works as an up-converter.

5.3.7 Bias Splitting Self-Oscillating Mixer (SOM)

The down-converted signal at the output of the VCO is measured at the sources of the transistors, M2 and M3, to avoid the degradation of the oscillator phase noise. As a result, the bias current generator M1 is divided into two transistors, M1a and M1b, as shown in Fig. 5.8. Here, the short circuit between M2 and M3 is substituted with a capacitor C that closes the loop at RF but provides high impedance at IF. This capacitor acts as a degenerative element for M2 and M3 and its value must satisfy

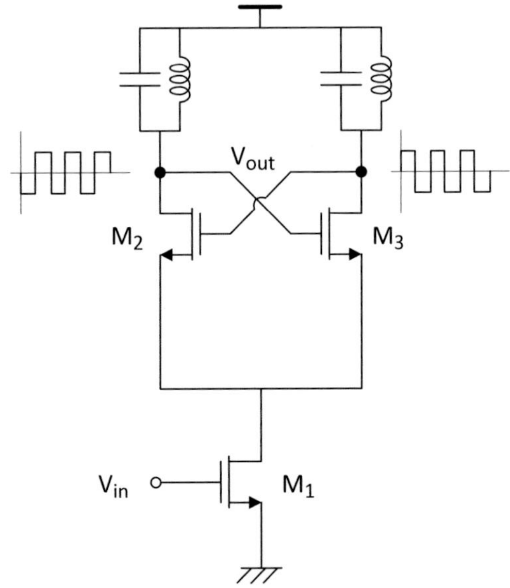

Fig. 5.7 LC tank oscillator as a mixer

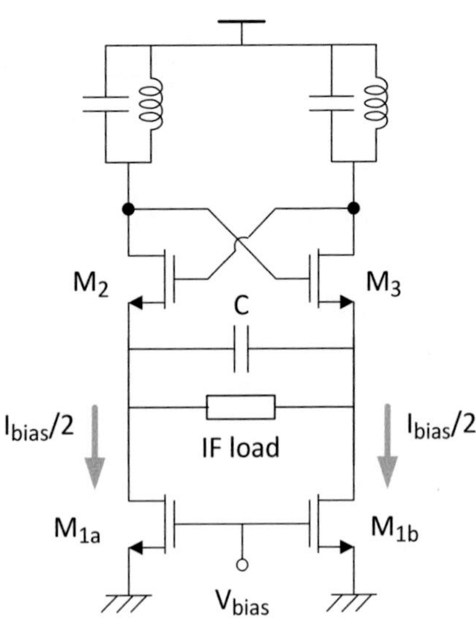

Fig. 5.8 Bias splitting circuit

$$\omega_{LO}C \gg g_{m2,3} \tag{5.10}$$

where $f_{LO} = 2\pi/\omega_{LO}$ is the oscillation frequency and $g_{m2,3}$ are the transconductance of M2 and M3.

The condition in Eq. 5.10 must be satisfied so that the loop gain is not reduced significantly. Otherwise, oscillation monotonically decreases and eventually stops.

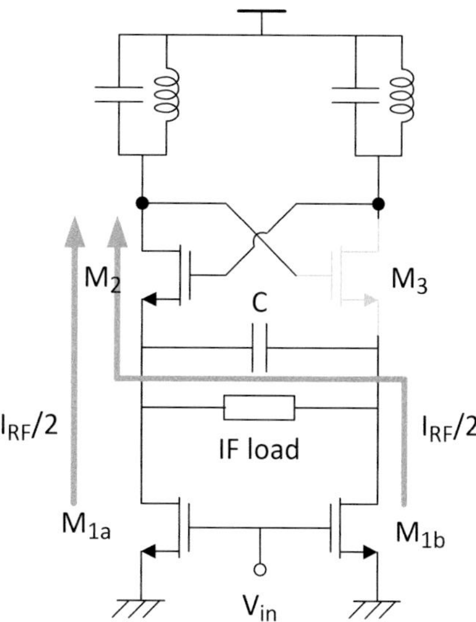

Fig. 5.9 Bias splitting SOM circuit

The modified bias splitting SOM structure is shown is Fig. 5.9 where the down-converted signal is measured at the sources of M2 and M3 across an IF load which may be the input impedance of a band-pass or a low-pass filter. During the first half of the LO period, M2 turns on and M3 turns off. As a result, the current, $I_{RF}/2$, injected by M1a flows directly into M2. Similarly, the current, $I_{RF}/2$, coming from M1b flows through the IF load into the M2. During the second half of the LO period, M3 turns on and M2 turns off. In this time interval, the current, $I_{RF}/2$, injected by M1a flows through the IF load into M3, and the current, $I_{RF}/2$, coming from M1b flows directly into M3. As a result, the total input current, $I_{RF}/2$, is multiplied by a square wave and measured at the output without disturbing the local oscillator. The IF load can be a high impedance load to generate voltage gain at the IF output or selected as a virtual ground load to short the sources of M2 and M3 at low frequencies. The ideal conversion gain of the bias splitting SOM is given by

$$\frac{I_{out}}{V_{in}} = \frac{1}{\pi} \tag{5.11}$$

This is because only half of the total RF current is injected by M1a and M1b, and the current alternatively flows through the IF load.

To double the conversion gain, a switching pair of SNTs, M4 and M5, can be added between M1 and the original switching pair, M2 and M3, as shown in Fig. 5.10. This structure is called double-switching pair SOM where M4 and M5 transistors are driven in opposite phases in contrast to M2-M3 pair. With M2 and M4 turned on simultaneously, all I_{RF} current flows through the IF load resulting in the conversion gain of

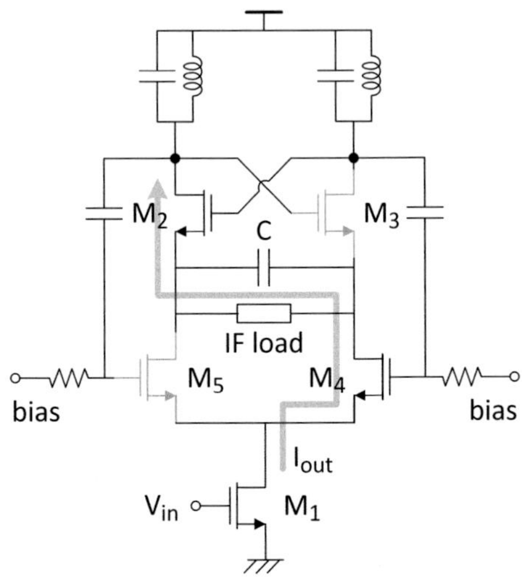

Fig. 5.10 Double-switching pair SOM circuit

$$\frac{I_{out}}{V_{in}} = \frac{2}{\pi} \tag{5.12}$$

Similarly, when M3 and M5 are turned on at the same time, the same conversion gain is achieved and all the current flows through the IF load. Because there are three transistors stacked on top of each other, there will be a need to slightly increase the supply voltage compared to the bias splitting SOM topology in order to keep all transistors operate properly.

In an RF receiver, the LNA can be implemented using double-switching pair SOM architecture and reusing M1 in an inductive degenerated topology. In this combined architecture, LNA, Mixer and VCO can be merged in a single cell where they share the bias and the devices as shown earlier in Fig. 5.6. Although there are three transistors stacked on top of each other, the minimum supply voltage can be as low as

$$V_{DDmin} = V_{th} + 3V_{ov} \tag{5.13}$$

where V_{th} is the threshold voltage and $V_{ov} = V_{GS} - V_{th}$ is the overdrive voltage.

In the LMV cell, M1 provides DC bias current to the VCO and acts as an LNA. Transistors, M2 and M3, perform the mixing operation and contribute to the VCO operation along with the capacitor C. The significant advantages of the LMV structure are the current reuse, performing multiple functions without any conflicts, reduction in device count and compatibility with a low supply voltage.

5.3.8 *Design of Double-Switching Self-Oscillating Degeneration LMV Cell Using SNTs*

The cross-coupled transistors consist of ten SNTs connected in parallel on each side. The supply voltage is 1 V and the operating frequency is 214 GHz. The VCO output oscillates from 0 to 2 V.

The mixer capacitor, C, is calculated to be 1 fF. The real component of Z_{in} is 238 Ω and the imaginary component is equal to −22 kΩ. To eliminate the capacitive impedance, an inductor can be added to the gate terminal of M1 in series. The calculated value of this inductor is 16.35 nH. At the input, a wideband matching network can also be added to match the standard terminating resistance of 50 Ω to 238 Ω in order to increase the bandwidth.

Figure 5.11 shows the LMV noise figure of 0.3 dB and is comparable to the performance of more advanced receivers. A low noise figure means that the signal is processed in the receiver effectively because the additive noise in the receiver is negligible compared to the amount of noise from the antenna.

The bandwidth at the input of the LMV (at a magnitude of 320 mV) is approximately equal to 3 GHz as shown in Fig. 5.12. The bandwidth is measured where |S11| is 0.3 V and this is where more than 90% of the signal power is transferred from the antenna to the LNA.

Figure 5.13 measures the gain of the LMV, which is approximately equal to −10.2 dB at 214 GHz where the impedance matching is achieved. The signal at the output is amplified to the maximum level and the majority of power is transferred. However, all out-of-band signals experience a small gain or even attenuation; their power from antenna is rejected and not delivered to receiver.

The LMV output signal spectrum is shown in Fig. 5.14. For a measured bandwidth of 3 GHz, the entire LMV down-converter operates linearly with no intermodulation distortion at the output.

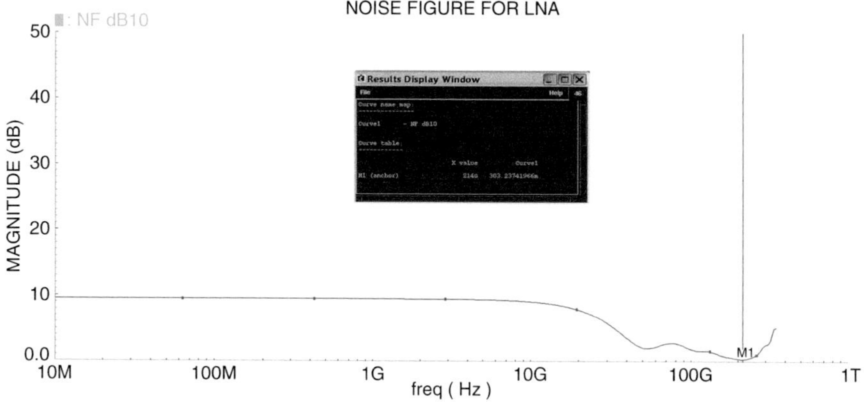

Fig. 5.11 The LMV noise figure

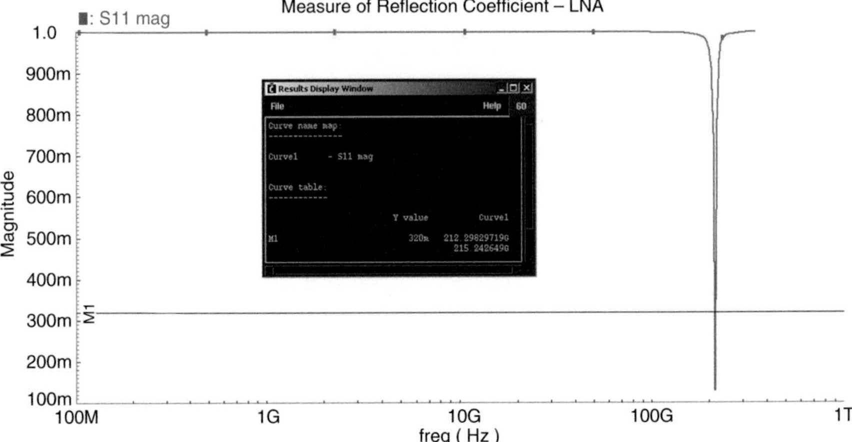

Fig. 5.12 LMV input impedance matching bandwidth

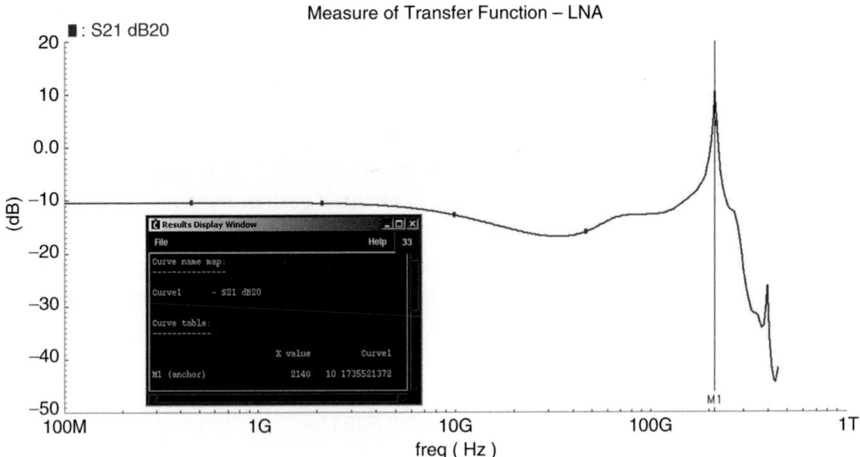

Fig. 5.13 The LMV S21

5.4 Variable Gain Amplifier (VGA)

5.4.1 Introduction to VGA

The Variable Gain Amplifier (VGA) is used in a variety of applications such as in Automatic Gain Control (AGC) circuits in wireless telecommunications. The objective in an application like this is to adjust the RF signal level accurately at the input of the VGA so that the baseband processor operates with correct data. If there are two receivers in a network at different distances from a transmitter, the

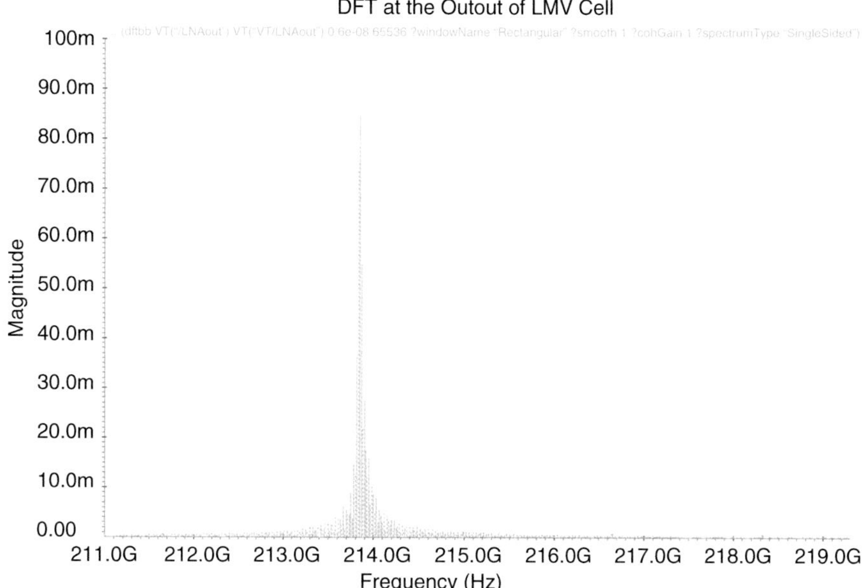

Fig. 5.14 The LMV RF spectrum

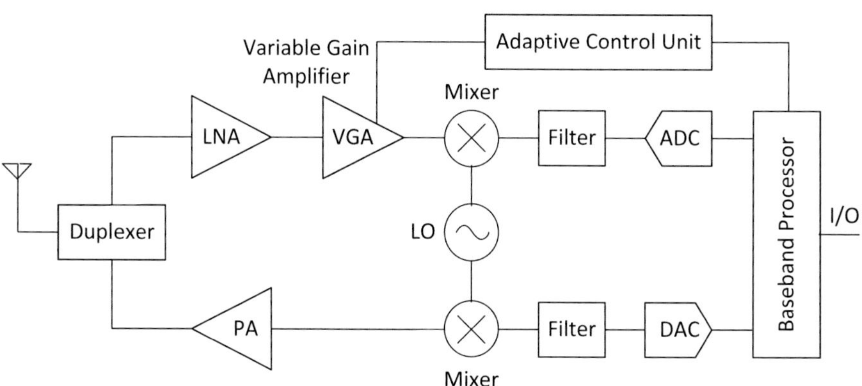

Fig. 5.15 Block diagram of the RF receiver with VGA

receiver with a shorter distance ends up receiving stronger signals that may saturate the receiver and generate incorrect data. Adding a variable gain amplifier attenuates the input signal and fixes the issue. If the baseband processor detects errors, it adaptively changes the gain of the VGA until error is minimized as shown in Fig. 5.15.

The VGA circuit presented in this chapter has an exponential response so that it is linear in decibel scale [4]. Typically, bipolar transistors, which have inherent exponential function built into their voltage response, are good candidates to use in VGA circuits. However, our objective in this chapter is to use SNTs [1] instead of bipolar transistors and optimize the VGA circuit so it represents the functionality of a Taylor series expansion of a function, f(x), or a pseudo-exponential equation [4].

$$f(x) = \left(\frac{1+x}{1-x}\right)^n \approx e^{2nx} \tag{5.14}$$

Here, $|x| < 1$.

5.4.2 Current-Mode Topology

A current-mode VGA is a variable gain amplifier that operates as a function of drain-source current. In a current-mode, exponential-control VGA circuit shown in Fig. 5.16 the output is approximately given by [4]

$$V_{out} = V_{bi}\exp\left(\frac{I_C}{I_B}\right) \tag{5.15}$$

Here, the gain is exponentially controlled by the current, I_C.

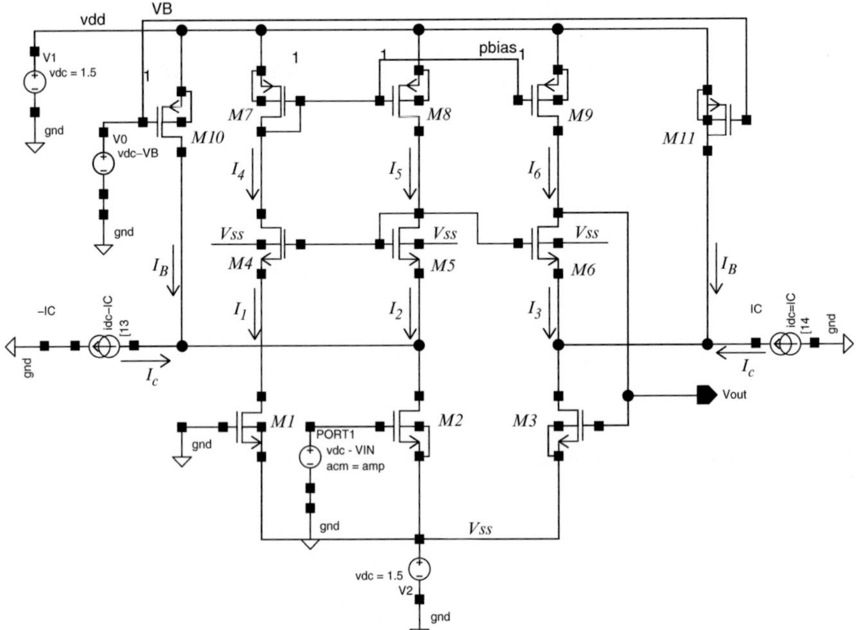

Fig. 5.16 Circuit diagram of the current-mode exponential-control VGA using SNTs

Table 5.1 Aspect ratios of the SNTs used in the simulation

Transistors	M1	M2	M3	M4	M5	M6	M7	M8	M9	M10	M11
Aspect ratio $\left(\frac{W}{L}\right)\left(\frac{nm}{nm}\right)$	$\frac{12.6}{10}$	$\frac{50.4}{10}$	$\frac{50.4}{10}$	$\frac{12.6}{10}$	$\frac{12.6}{10}$	$\frac{12.6}{10}$	$\frac{100.8}{10}$	$\frac{100.8}{10}$	$\frac{100.8}{10}$	$\frac{126}{10}$	$\frac{126}{10}$

Table 5.2 The equivalent number of parallel SNTs in Table 5.1

Transistors	M1	M2	M3	M4	M5	M6	M7	M8	M9	M10	M11
Number of transistors in parallel	1	4	4	1	1	1	8	8	8	10	10

The SNT aspect ratios used in simulations are shown in Table 5.1. The number of transistors connected in parallel is shown in Table 5.2 in order to satisfy the transistor sizes in Table 5.1.

Transistors, M10 and M11, are chosen to have width to length ratio, W/L, of 126/10 nm, to provide $I_B = 30$ μA when $I_C = 0$ A at $V_{in} = 500$ mV and $V_B = 970$ mV as shown in Fig. 5.16. The value of V_B is chosen to maximize g_m of M10 and M11, and I_B of M10 and M11, which is dependent on the V_{ds} values of M2 and M3. To compensate this, the size of M_2 and M_3 can be adjusted. For the final design in Table 5.1, setting V_{in} to 0 V makes $I_C = 0$ A. In this case, the main currents flowing through the circuit become $I_1 = 12.3$ μA, $I_5 = 15$ μA, $I_6 = 15.5$ μA, and $I_4 = 12.3$ μA.

We vary I_C and measure the change in V_{out} as I_C can impact both the DC and AC components of V_{out}. Here, it is important to verify if V_{out} can be changed exponentially with I_C and plotted as a function of I_C. The DC output voltage response of the current-mode exponential-control VGA is shown in Fig. 5.17a. The dynamic range for the DC output voltage response is approximately 9 dB. The AC output voltage response is shown in Fig. 5.17b. The dynamic range for the AC output voltage response is approximately between 50 dB and 53 dB for 500 mV $\leq V_{IN} \leq$ 700 mV. The semi-linear response of V_{out} in Fig. 5.17a, b shows that the VGA is exponentially controlled.

The AC gain and phase of the current-mode VGA circuit is shown in Fig. 5.18. The gain of the amplifier is flat and with a bandwidth of 327 GHz. Changing I_C in 5 μA increments increases the gain in 6 dB. The amplifier shows stability because its gain is less than 0 dB and its phase is less than 100° within the 327 GHz bandwidth.

As we pointed out earlier, the VGA adjusts the signal to a desired level not to saturate the amplifier. If the signal at receiver input is weak, understanding the embedded noise component of the signal becomes very important. If the receiver block generates large amount of noise, the signal may be buried inside the noise and becomes impossible to detect. As a result, the low signal-to-noise ratio of the VGA may deteriorate the performance of the entire receiver. The input-referred noise of the current-mode VGA is shown in Fig. 5.19. The input noise is linear from DC to 84 GHz. For $I_C = -5$ μA, the input noise is approximately 25 nV/Hz$^{1/2}$. Similarly, for $I_C = 0$ μA, the input noise reaches approximately 40 nV/Hz$^{1/2}$ and for $I_C = 5$ μA,

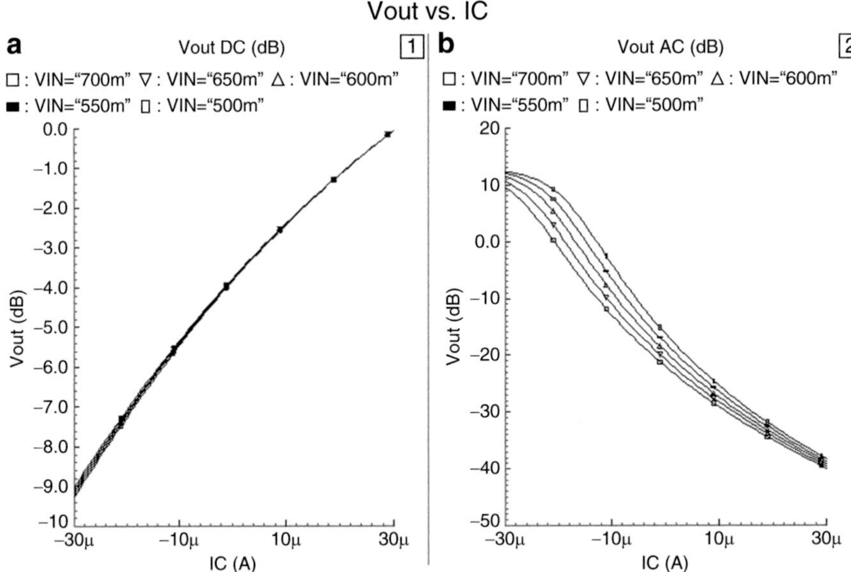

Fig. 5.17 (**a**) DC and (**b**) AC output voltage response of the current-mode VGA in dB for 500 mV < V_{IN} < 700 mV

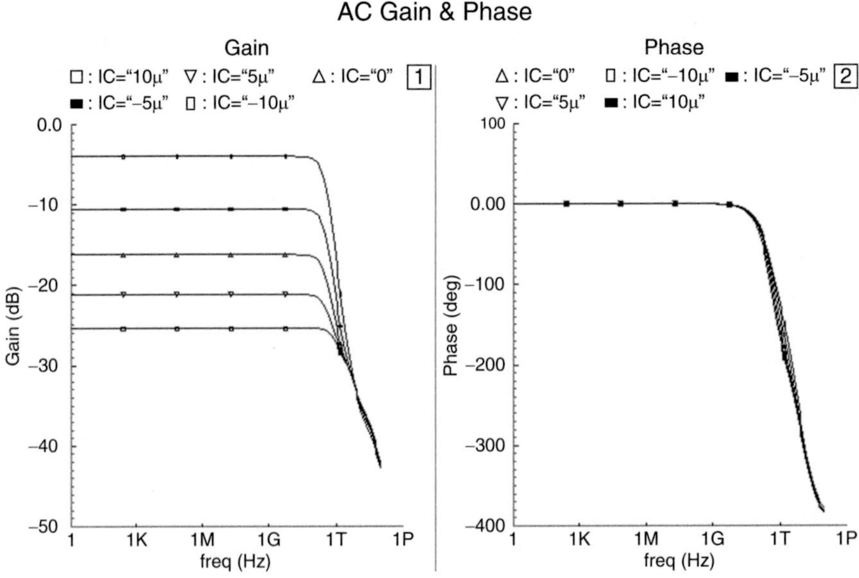

Fig. 5.18 AC gain and phase of the current-mode VGA

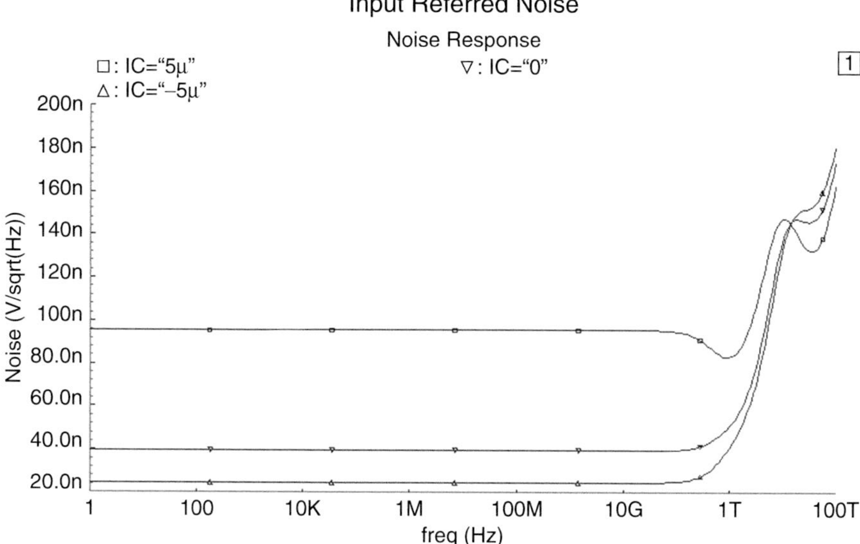

Fig. 5.19 Input-referred noise of the current-mode VGA

it becomes approximately 95 nV/Hz$^{1/2}$. Therefore, a receiver using the current-mode VGA configuration is easily capable of handling extremely weak signals in range of several hundred nanovolts.

Sometimes the noise in the amplifier output may be higher or lower than the noise at its input. Therefore, it is helpful to study the output noise of the current-mode VGA and compare it against its input noise. The output-referred noise is shown in Fig. 5.20. The output noise is linear from DC to 70 GHz. For $I_C = -5$ μA, the output noise becomes equal to 970 pV/Hz$^{1/2}$; for $I_C = 0$ μA and $I_C = 5$ μA, the output noise reaches as high as 961 pV/Hz$^{1/2}$ and 960 pV/Hz$^{1/2}$, respectively. For the current-mode VGA, both the signal and the noise are varied in proportion to each other. Therefore, maintaining an acceptable signal-to-noise ratio at the input produces an acceptable signal-to-noise ratio at the output of the amplifier and ensures proper operation of the receiver.

5.4.3 Voltage-Mode Topology

An exponentially controlled voltage-mode VGA circuit is shown in Fig. 5.21. The gates of the transistors, M10 and M11, are connected to the voltage levels, $(V_B + V_C)$ and $(V_B - V_C)$, respectively. Here, V_B is a bias voltage and V_C is a control voltage for the amplifier. Assuming that M10 and M11 are perfectly matched transistors satisfying $V_B < 0$ V and $(V_{DD} - V_B) > |V_{TP}|$, the output is approximately given by [4].

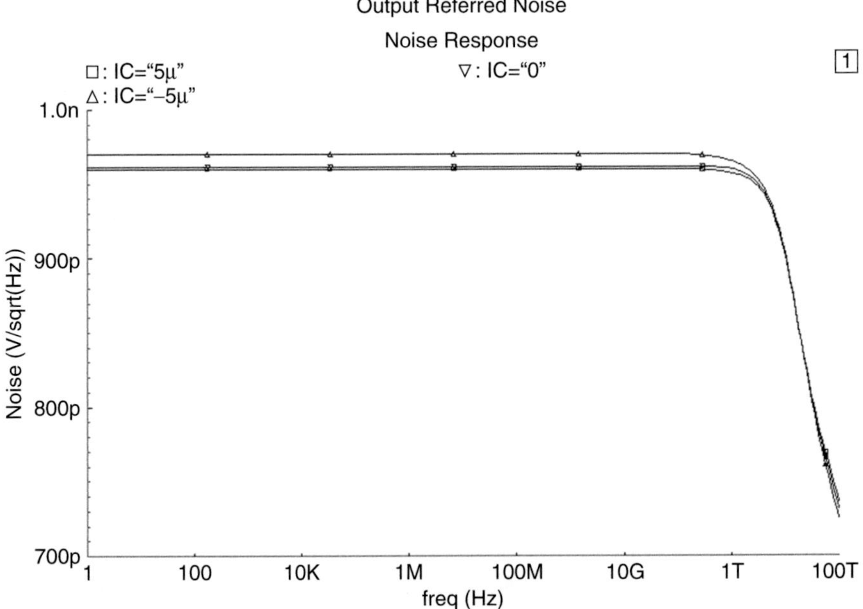

Fig. 5.20 Output-referred noise of the current-mode VGA

$$V_{out} \approx V_{in} \exp\left(4 \frac{V_C}{V_{DD} - V_B - |V_{TP}|}\right) \tag{5.16}$$

Here, V_{in} is the voltage applied to the gate of M2, V_B is a bias voltage, V_C is the control voltage applied to gates of M10 and M11, and V_{TP} is the threshold voltage of M10 and M11. According to this equation, the gain of a VGA can be exponentially controlled by the voltage V_C. The schematic used to simulate the voltage-mode exponential-control VGA is shown in Fig. 5.21.

The SNT aspect ratios for transistors used during simulation are listed in Table 5.3.

In Fig. 5.21, V_B is mainly used to bias transistors in order to operate them at the desired linear region. For a proper circuit operation, it is important that V_{out} varies exponentially if V_C is changed linearly as V_C can impact both DC and AC components of V_{out}. The DC output voltage response of the voltage-mode, exponential-control VGA is shown in Fig. 5.22a. The dynamic range for the DC output voltage response is approximately 6.5 dB. The AC output voltage response is shown in Fig. 5.22b. The dynamic range for the AC output voltage response is approximately between 35 dB and 40 dB for 350 mV $\leq V_{IN} \leq$ 500 mV.

The AC gain and phase response of the voltage-mode, exponential-control VGA is shown in Fig. 5.23. The gain response is flat and has a bandwidth of 311 GHz. Changing I_C in 5 μA increments increases the gain in 8 dB steps. The AC gain shown in Fig. 5.23 implies that the gain increases exponentially with a constant rate.

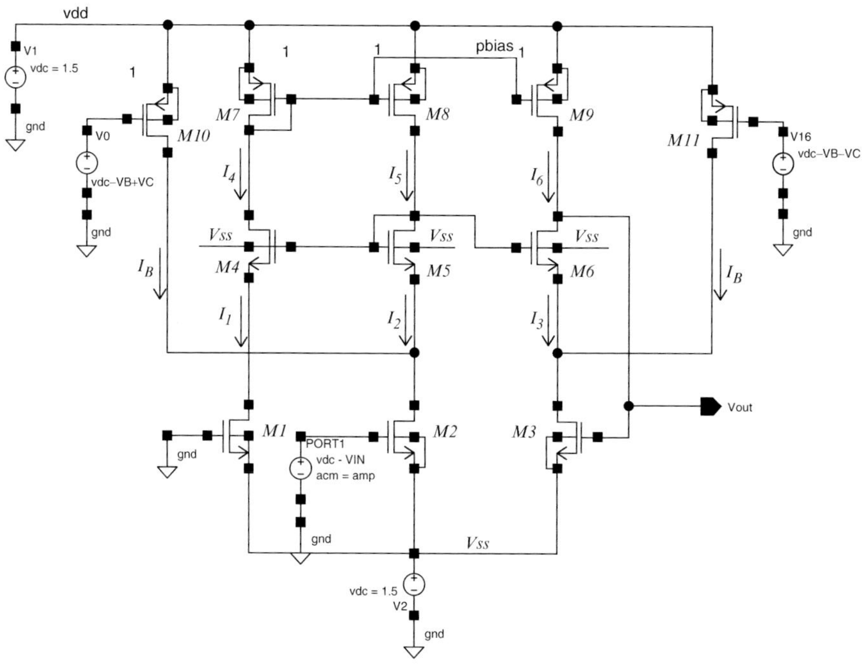

Fig. 5.21 Voltage-mode exponential-control VGA using SNT process

Table 5.3 Aspect ratios of the SNTs used in the simulation

Transistors	M1	M2	M3	M4	M5	M6	M7	M8	M9	M10	M11
Aspect ratio $\left(\frac{W}{L}\right)\left(\frac{nm}{nm}\right)$	$\frac{12.6}{10}$	$\frac{50.4}{10}$	$\frac{50.4}{10}$	$\frac{12.6}{10}$	$\frac{12.6}{10}$	$\frac{12.6}{10}$	$\frac{100.8}{10}$	$\frac{100.8}{10}$	$\frac{100.8}{10}$	$\frac{126}{10}$	$\frac{126}{10}$

Also, the AC gain and the phase in this figure confirm that the circuit is stable because the phase is less than 100° for all frequencies between 0 and 311 GHz.

The input-referred noise is shown in Fig. 5.24. The input noise is linear from 0 Hz to 28 GHz. For $V_C = -50$ mV, the input noise is equal to 20 nV/Hz$^{1/2}$. For $V_C = 0$ mV and $V_C = 50$ mV, the input noise becomes approximately 40 nV/Hz$^{1/2}$ and 294 nV/Hz$^{1/2}$, respectively.

The output-referred noise is shown in Fig. 5.25. The output noise is linear between 0 Hz and 42 GHz. For $V_C = -50$ mV the output noise is approximately 981 pV/Hz$^{1/2}$. Similarly, for $V_C = 0$ mV and $V_C = 50$ mV, the output noise reaches 961 pV/Hz$^{1/2}$.

The characteristics of the two VGA topologies studied in this chapter are compared in Table 5.4.

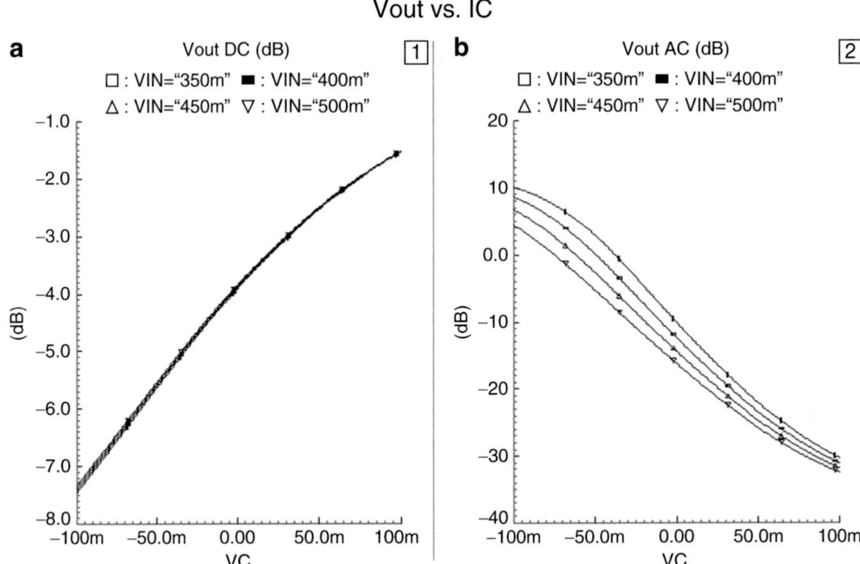

Fig. 5.22 (**a**) DC and (**b**) AC output voltage response of the voltage-mode VGA in dB for 350 mV < V_{IN} < 500 mV

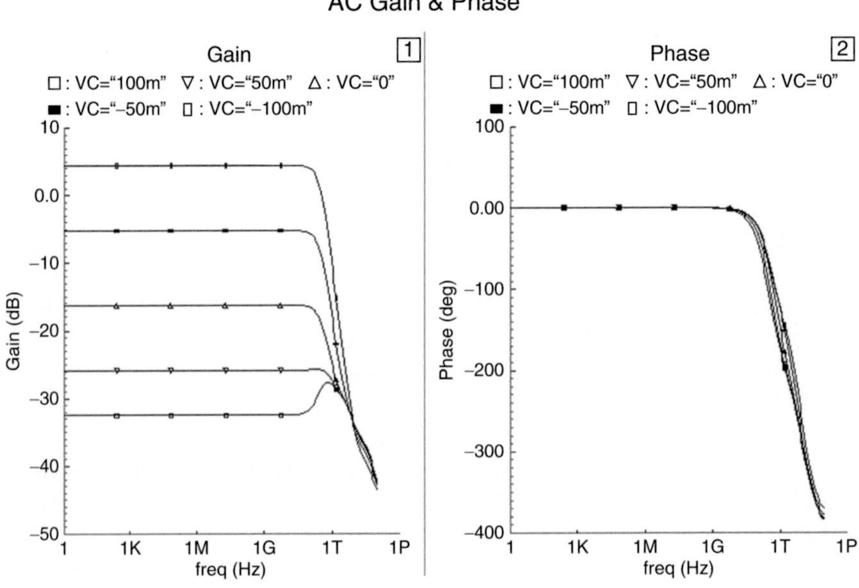

Fig. 5.23 AC gain and phase of the voltage-mode VGA

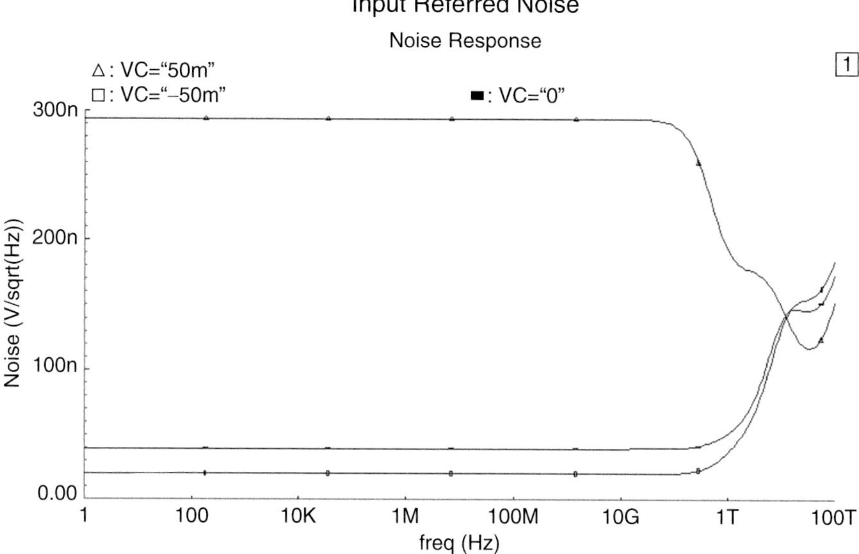

Fig. 5.24 Input-referred noise of the voltage-mode VGA

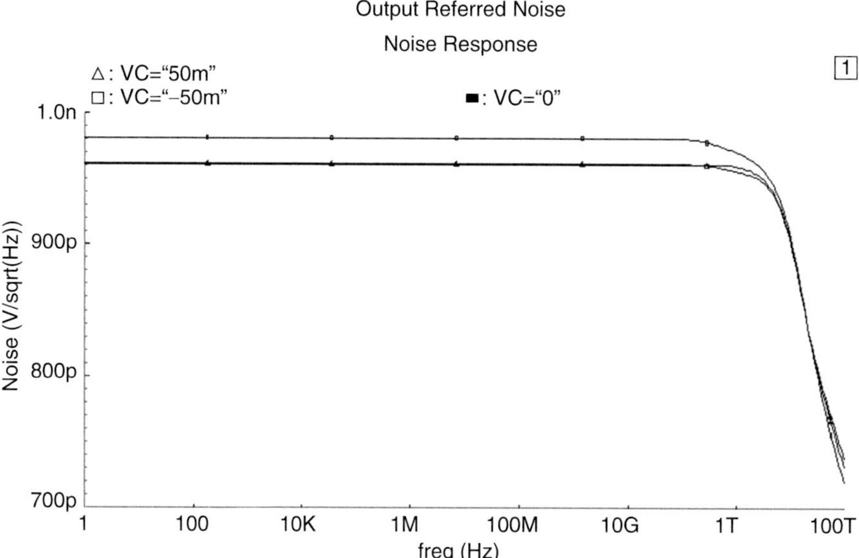

Fig. 5.25 Output-referred noise of the voltage-mode VGA

Table 5.4 Characteristics of the two VGA topologies

Parameter	Current mode	Voltage mode
Supply voltage	1.5 V	1.5 V
Minimum gain	−38.5 dB	−33.4 dB
Maximum gain	12.6 dB	10.0 dB
−3 dB frequency	327 GHz at $I_c = -5$ μA	311.3 GHz at $V_c = -0.05$ V
	373 GHz at $I_c = 0$ μA	375.7 GHz at $I_c = 0$ μA
	528 GHz at $I_c = 5$ μA	1.832 THz at $V_c = 0.05$ V
Input-referred noise	25 nV/Hz$^{1/2}$ at $I_c = -5$ μA	20 nV/Hz$^{1/2}$ = at $V_c = -0.05$ V
	40 nV/Hz$^{1/2}$ at $I_c = 0$ μA	40 nV/z$^{1/2}$ at $V_c = 0$ V
	95 nVHz$^{1/2}$ at $I_c = 5$ μA	294 nV/H$^{1/2}$ at $V_c = 0.05$ V
Power consumption	0.33 mW	0.33 mW

5.5 Summary

Wireless transceivers operating above 100 GHz are becoming very popular because they offer fast data rates suitable for big data storage and multichannel, high definition video generation. All circuits studied in this chapter operate well above 100 GHz and are good candidates for modern communication systems. Circuits implemented with SNTs show a lower noise compared to the circuits built by planar MOS transistor technologies where the flicker noise contributes to high noise values. Furthermore, RF circuits designed with SNTs have extremely low power consumption. Low-power wireless receivers (and transmitters) using this technology can be used to implement Internet of Things (IOT) where the power constraints are extremely crucial.

The first RF block studied in this chapter is the LMV circuit consisting of the LC tank oscillator, the mixer and the LNA. These three functional units can be combined together to generate a down-converter prior to a baseband processor. The LNA can be connected to a self-oscillating mixer structure which contains a double crossing mixer and a VCO. The VCO oscillation frequency is measured to be 214 GHz. The LMV cell has the noise figure of 0.3 dB and an approximate bandwidth of 3 GHz. The total RMS power dissipation is approximately 20 μW at a 1 V power supply.

The second block studied in this chapter is the variable gain amplifier. VGAs are used in communication systems where the distance change between a transmitter and a receiver or environmental variations will strongly impact the signal strength at the receiver. Adding a VGA to the system can also compensate for signal strength loss and ensures proper receiver operation. Two new exponentially controlled VGA circuits presented in this chapter are current-mode and voltage-mode amplifiers. The current-mode VGA operating at 1.5 V supply voltage produces a variable gain from −38.5 dB to 12.6 dB, a bandwidth above 327 GHz, and a noise figure less than 95 nV/Hz$^{1/2}$. The voltage-mode VGA operating at 1.5 V supply voltage provides a variable gain from −33.4 dB to 10 dB, a bandwidth above 311 GHz, and a noise figure less than 95 $nV/Hz^{1/2}$. Both topologies have a total power dissipation of approximately 0.33 mW.

References

1. Hamedi-Hagh S, Bindal A (2008) SPICE modeling of silicon nanowire field effect transistors for high speed analog integrated circuits. IEEE Trans Nanotechnol 7:766–775
2. Hamedi-Hagh S, Bindal A (2008) Characterization of nanowire CMOS amplifiers using fully depleted surrounding gate transistors. J Nanoelectron Optoelectron 3(3):281–288
3. Liscidini A, Mazzanti A, Tonietto R, Vandi L, Andreani P, Castello R (2006) Single-stage low-power quadrature RF receiver front-end: the LVM cell. IEEE J Solid-State Circuit Conf 41:2832–2841
4. Liu W, Liu S-I, Wei S-K (2004) CMOS exponential-control variable gain amplifiers. IEEE Proc Circuits Devices Syst 151(2)

Chapter 6
SRAM Mega Cell Design for Digital Applications

6.1 Introduction

The primary objective of this chapter is to present the application of silicon nanowire technology on a first large-scale digital mega cell design: an SRAM. The detailed steps of generating accurate BSIMSOI SPICE models from vertically-grown SNTs with undoped bodies and dual work function metal gates were already discussed in Chapter 3. This chapter uses the SNT device models in the design and analysis of a 16×16 SRAM block and reports the circuit simulation results and electrical data.

6.2 Brief Description of Transistor Design and Modeling

Both NMOS and PMOS transistors used in this chapter are enhancement type with undoped, cylindrical silicon bodies constructed perpendicular to the substrate as shown in Chapter 1 and then resumed in Chapter 3. Source/Drain (S/D) contacts are assumed to have ohmic contacts and 1.5 nm thick gate oxide. Device simulations are performed using Silvaco's three-dimensional ATLAS device simulation environment with a 1 V power supply voltage. Half of the device is constructed in a two-dimensional platform and then rotated around the y-axis to create a three-dimensional cylindrical form for simulations. The device radius is changed from 2 to 20 nm while its effective channel length is varied between 10 and 65 nm. The device simulator used low and high-electric field mobility models, concentration dependent Shockley-Read-Hall recombination model, Arora's lattice temperature model, Selberherr's impact ionization model, and Fermi statistics. Quantum mechanical effects are included using density gradient method.

The device design process starts by determining individual metal gate work functions for each NMOS and PMOS SNT that produces 300 mV threshold voltage.

A. Bindal, S. Hamedi-Hagh, *Silicon Nanowire Transistors*,
DOI 10.1007/978-3-319-27177-4_6

Once the gate work function for each transistor is determined, the body radius and effective channel length of both SNTs are simultaneously changed until each device reveals minimum static and dynamic power dissipations but exhibits the fastest transient times. This design process has ultimately produced an SNT body of 2 nm radius and 10 nm channel length. The BSIMSOI device models are based on these particular device dimensions and they given in Chapter 3 and they are used for all the circuit simulations in this chapter.

6.3 SRAM Design

This section demonstrates the overall SRAM architecture including the core, address decoder, read/write data-path and the self-time circuits. Power dissipation as well as read and write access times under different ambient conditions are also discussed.

6.3.1 SRAM Architecture

The SRAM architecture in this chapter consists of an SRAM core for data storage and retrieval, self-timed circuits for controlling read and write operation sequences, and an address decoder to decode a 4-bit wide address in order to produce 16 Word Lines (WL) as shown in Fig. 6.1.

To write into the SRAM block, WrEn is set high while RdEn is set low. This precharges the SRAM core for a write operation, enables the address decoder, and validates the input data. To read from the SRAM block, WrEn is set low and RdEn is set high. This setting enables the precharging circuit and the address decoder, but it also produces a strobe pulse to activate the sense amplifier to speed up the read operation. When both WrEn and RdEn are set low, no precharge pulse is produced to conserve power; the input data and address become invalid; only the prior output can be read from DataOut port.

6.3.2 SRAM Core

The 16×16 SRAM core consists of 16 identical columns, each of which includes 16 rows of six-transistor (6 T) memory cells for bit storage, a precharging circuit, a sense amplifier for read operation, a write circuit, and an output latch/buffer as shown in Fig. 6.2.

The input–output characteristics of the 6 T memory cell are shown in Fig. 6.3. The inverter threshold voltage is approximately 470 mV even though the ON

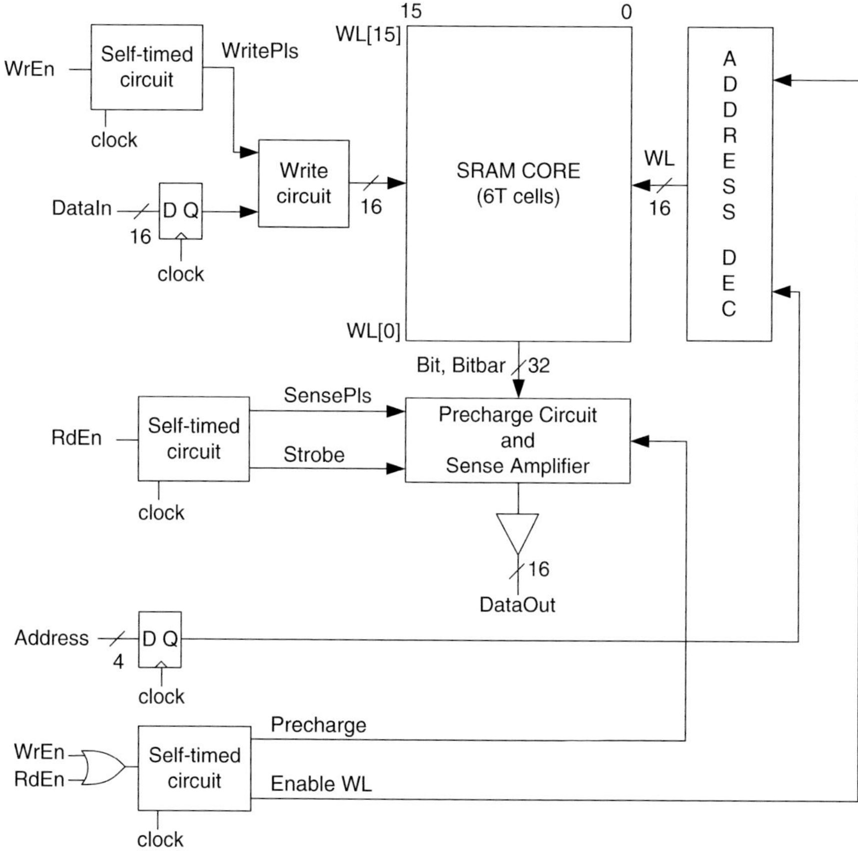

Fig. 6.1 16 × 16 SRAM architecture

current value of the NMOS SNT is twice as large as the ON current of the PMOS SNT. The low and high noise margin figures are 360 mV and 440 mV, respectively.

A typical write operation is initiated with a high WrEn signal which produces an active-low precharge pulse immediately after the positive edge of clock as shown in Fig. 6.4. Note that the positive clock edge corresponds to the origin of each graph. Figure 6.4 does not represent the actual waveforms obtained during a write operation; however, it shows the complete write sequence and data validation checks under nominal conditions. During precharge pulse, both Bit and Bitbar lines are pulled up to 1 V prior to accessing a memory cell. The voltage level on Sense and Sensebar nodes is immaterial for a write operation since the PMOS transistors, S1 and S2, are turned off. The active-low WritePulse and the active-high EnableWL signals must be generated following the input data and address for validation, respectively. The combination of EnableWL and valid address generates a WL pulse which turns on the pass-gate transistors, N1 and N2, to access a 6 T cell.

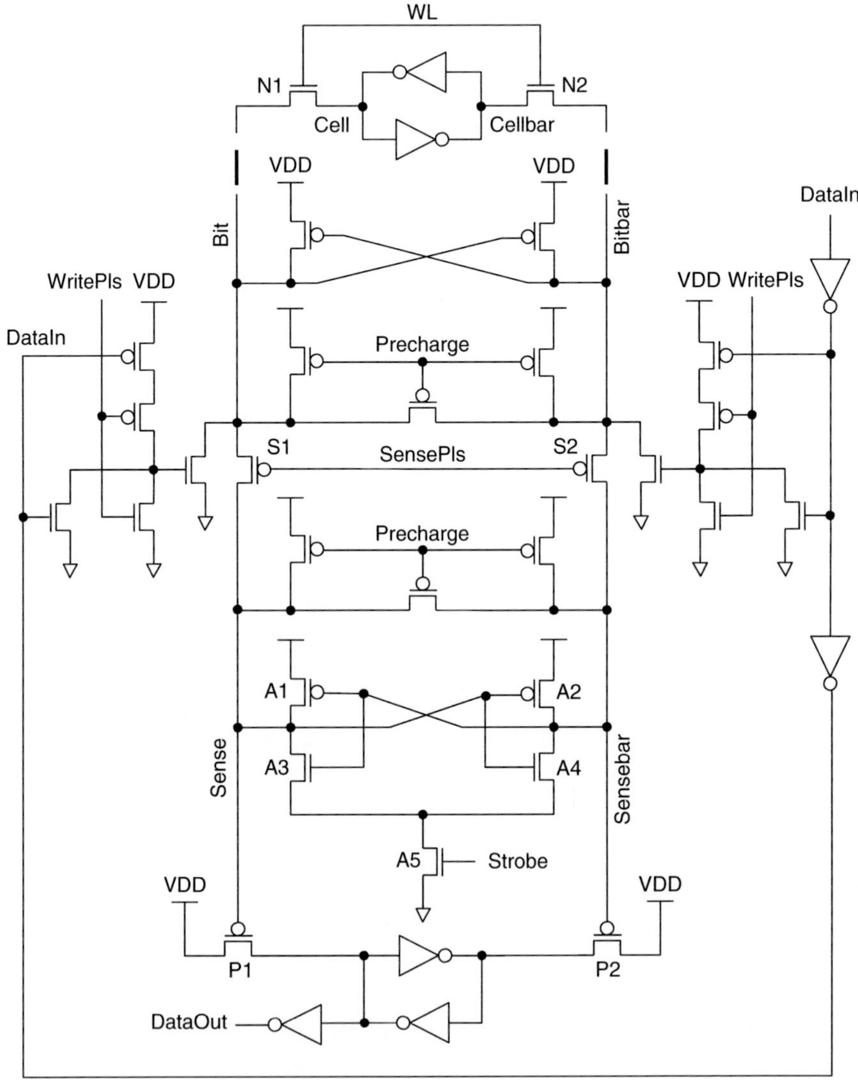

Fig. 6.2 An SRAM column consisting of 6 T cell, precharge, read and write circuits, and output buffer

While WL is active, the valid input data is written into the cell. WritePulse is terminated as soon as the input data is latched in the memory cell which alters the voltage levels at Cell and Cellbar nodes.

A typical read operation is initiated by a high RdEn signal which also produces a precharge pulse to pull up Bit, Bitbar, Sense and Sensebar lines to 1 V as shown in Fig. 6.5. Again, this graph does not show actual waveforms but reveals the complete

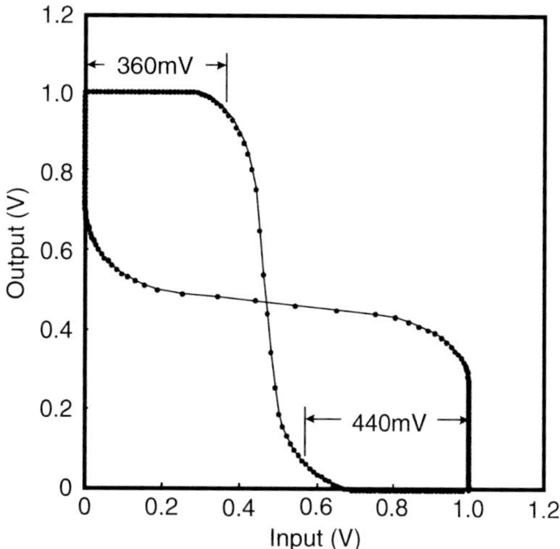

Fig. 6.3 Input–output characteristics of a 6-transistor (6 T) memory cell

read sequence and data validation checks under nominal conditions. Identical to the write operation, the active-high EnableWL signal is generated to validate an input address and turn on the pass-gate transistors of the memory cell. When the differential voltage between the Bit and Bitbar nodes reaches approximately 50 mV, the Strobe pulse is produced to activate the sense amplifier. During this period, the active-low SensePulse signal is also generated to turn on the transistors, S1 and S2, to transfer the charge from the Bit (Bitbar) node to the Sense (Sensebar) node. The ratio of the capacitance at the Bit (Bitbar) and Sense (Sensebar) nodes determines the initial voltage level at the Sense (Sensebar). The sense amplifier utilizes the higher transconductance values of the transistors, A1 and A4 (rather than A2 and A3), and pulls down the Sensebar node towards 0 V while sustaining 1 V at the Sense node. Consequently, P2 transistor turns on while P1 transistor stays off, allowing the LatchInbar and the DataOut nodes to reach 1 V. The Strobe and WL pulses are terminated when the Sensebar node reaches approximately 0 V and the contents of the memory cell are successfully stored at the LatchIn and Latchinbar nodes, respectively.

6.3.3 *Address Decoder*

The address decoder generates 16 WL signals for each row of the SRAM core. When enabled by the EnableWL, the valid 4-bit input address is decoded to activate only one row of the SRAM core. A disabled address decoder produces 0 V to all of its outputs. Subsequently, none of the rows is turned on for either read or write operation.

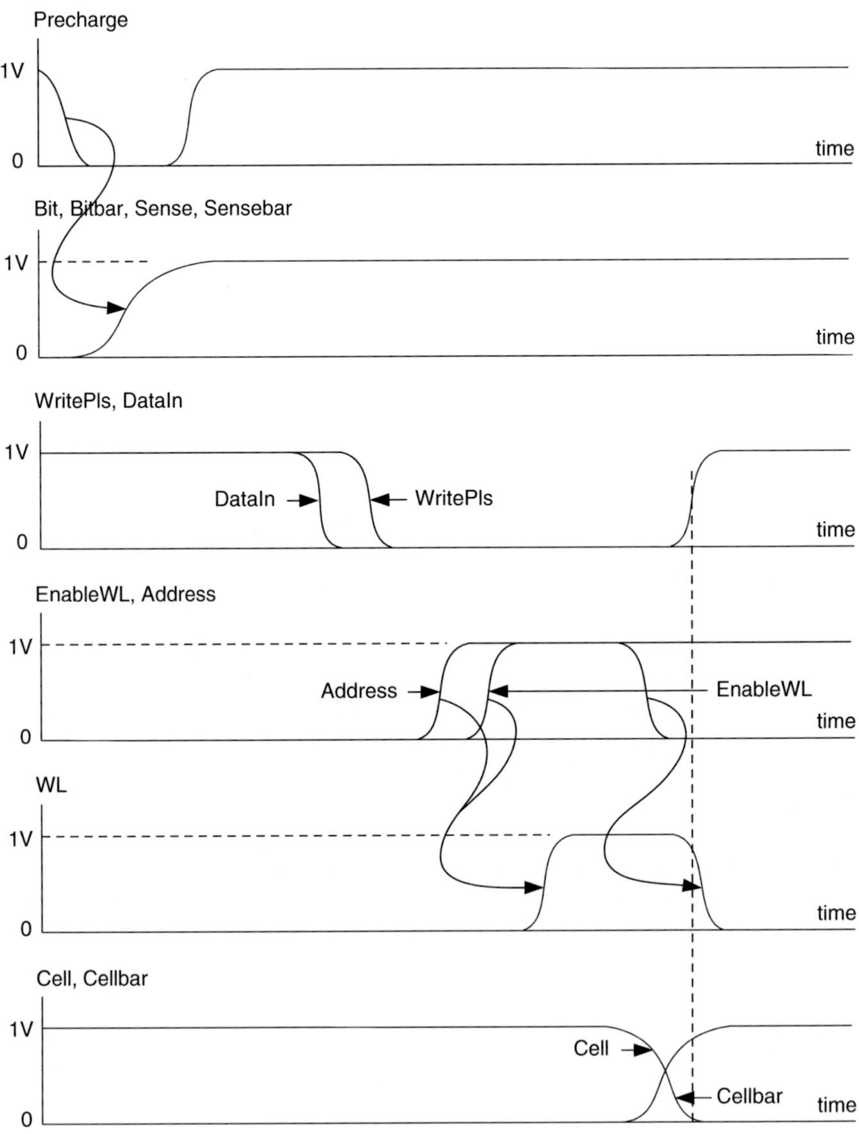

Fig. 6.4 A typical WRITE sequence

6.3.4 Self-Timed Circuits

Self-timed circuits produce either active-high or active-low signals for controlling the data flow sequence during a read or write operation. Figure 6.6 shows a typical self-timed circuit that generates an active-high pulse. Both the origin and duration of the pulse can be determined with respect to the positive edge of clock by isolated

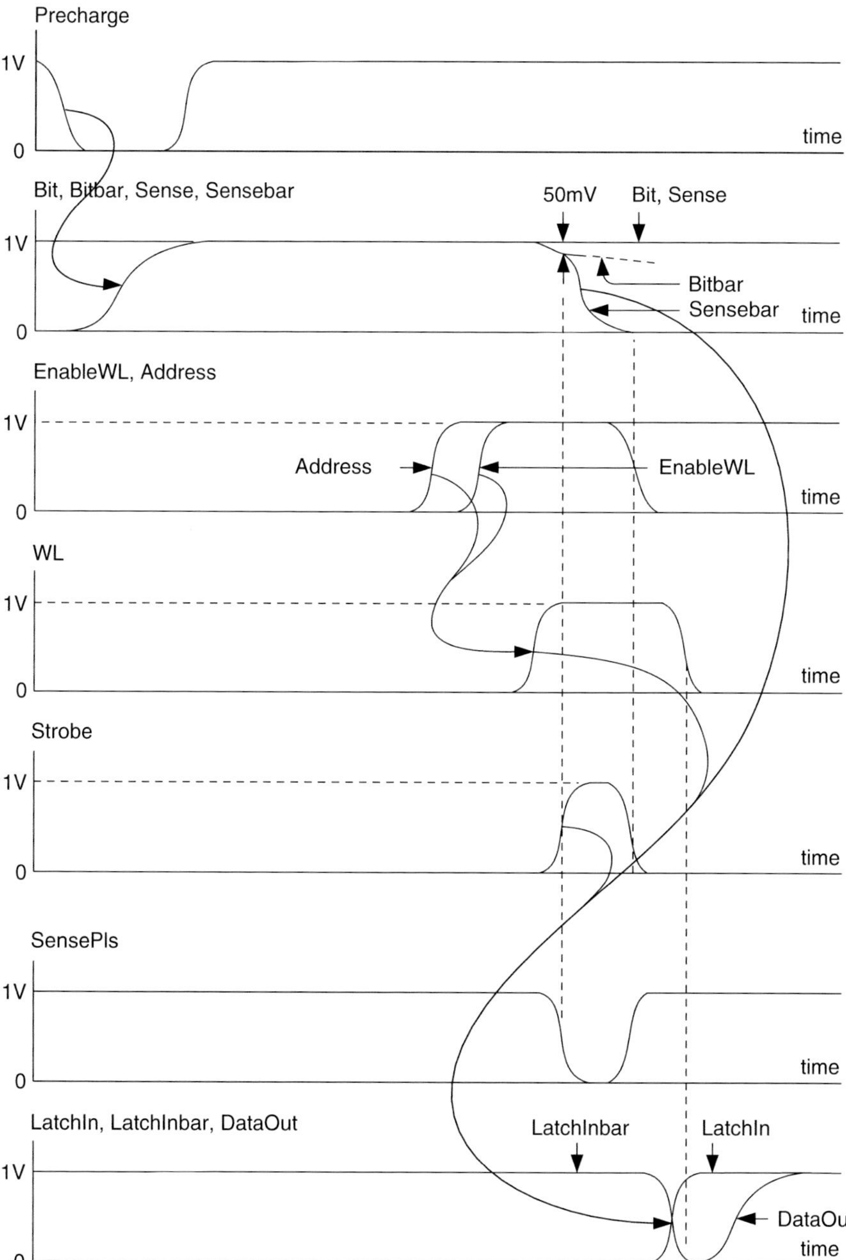

Fig. 6.5 A typical READ sequence

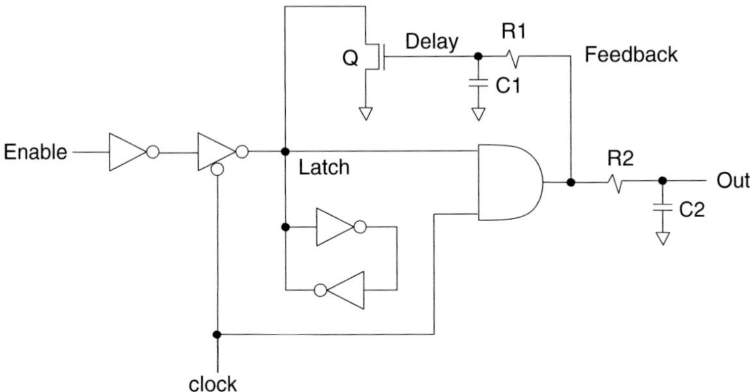

Fig. 6.6 Typical self-timed circuit producing an active-high pulse with respect to positive clock edge

Table 6.1 Time constant values of self-timed circuits

	R1 (MΩ)	C1 (aF)	R2 (MΩ)	C2 (aF)
Precharge	1.0	4.3	0.0	0.0
Enable	5.0	4.3	4.2	4.3
SensePls	9.2	4.3	7.8	4.3
Strobe	9.5	4.3	8.2	4.3
WritePls	4.0	4.3	2.6	4.3

RC circuits. An active-high Enable signal is latched in the self-timed circuit when clock is low. Since both the Feedback and Delay nodes are at 0 V, the NMOS transistor does not turn on; the voltage level at the Latch stays undisturbed at 1 V. However, as soon as clock goes high, the input tri-state inverter isolates the Latch node from the changes at the Enable input. The Feedback node goes high and pulls up the Delay node after a delay determined by R1 and C1. The NMOS transistor, Q, turns on and discharges the Latch node. Following this discharge, the Feedback node goes low and turns off Q after the same RC delay. The pulse generated at the Feedback node is reproduced at the Out node after a delay determined by R2 and C2. Therefore, R1 and C1 adjust the pulse duration while R2 and C2 establish the origin of this pulse with respect to the positive edge of clock. The values of R1, C1, R2 and C2 are listed in Table 6.1 for each self-timed circuit used in the SRAM block. The waveforms of the self-timed circuit at each critical node are shown in Fig. 6.7.

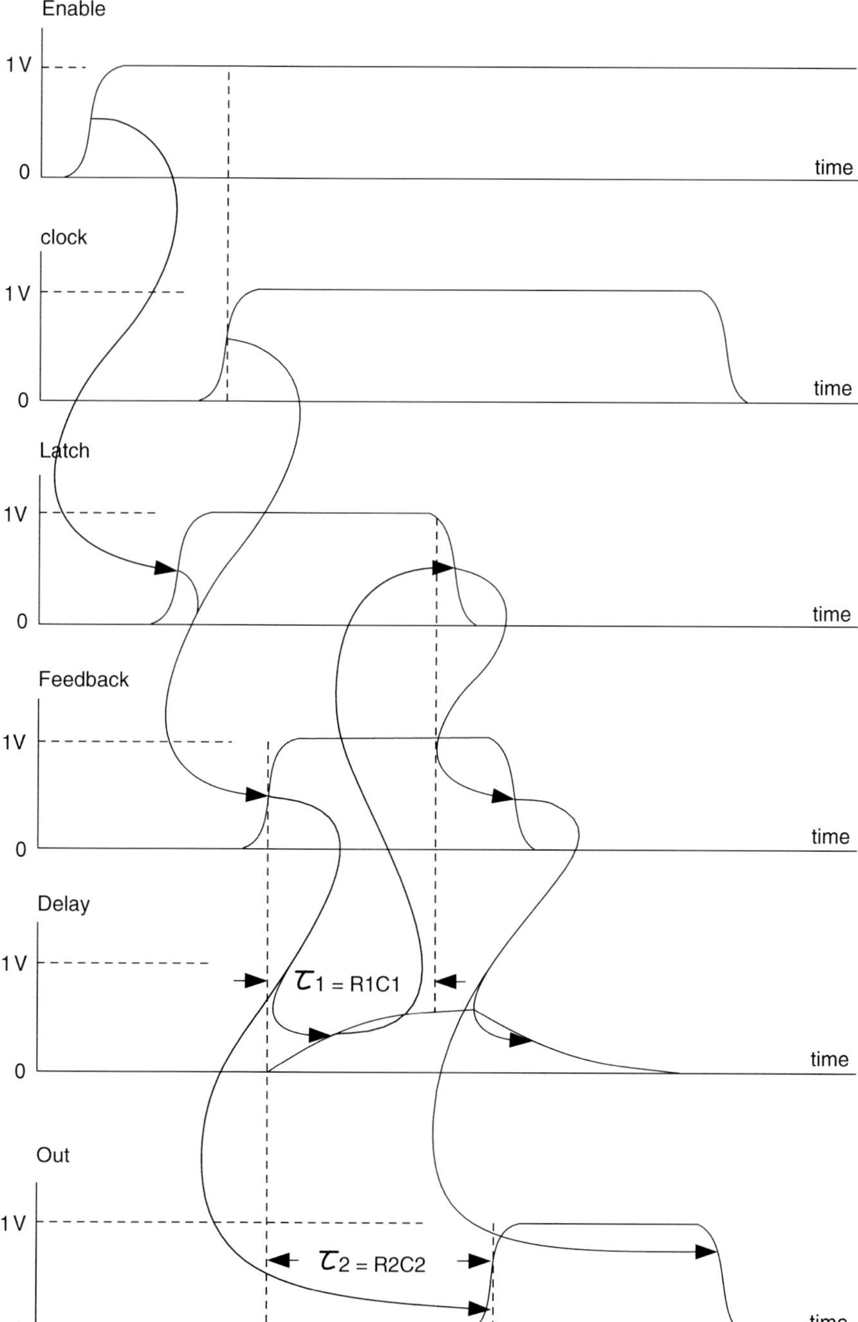

Fig. 6.7 Waveforms of a self-timed circuit

6.4 SRAM Characteristics

6.4.1 Parasitic Layout Extraction

6 nm wide and 1.4 aspect ratio (wire height to width) copper wires are used for interconnects. Since sub-10 nm range copper wire electrical characteristics do not exist in the literature, copper resistivity was extrapolated from Srivastava's model for 1.4 aspect ratio wires as discussed in Chapter 1, and 20 μΩ-cm resistivity was subsequently used to calculate the sheet resistance for 6 nm wide interconnects. Similarly, contact resistance was extrapolated from experimental data on 100 nm and larger via diameters and resulted in 18 Ω for each metal contact.

N-well and P-well contacts are formed with guard ring structures around each well periphery to reduce the NMOS and PMOS source extension resistances as discussed in Chapter 3. This scheme produces 2.3 kΩ N-well and 3.4 kΩ P-well extension resistances in series with the source terminal. Even though source extension resistance normally dominates over the combination of local interconnect and contact resistances, its magnitude is much smaller than the transistor channel resistance. The NMOS and PMOS equivalent channel resistances extracted from inverter rise and fall times are approximately 51.3 kΩ and 78.6 kΩ, respectively. Each resistance is approximately 22 times higher than the corresponding source extension resistance. The degradation in ON current due to source resistance becomes better than 8 % for both transistors. However, the guard ring structure increases the single transistor layout area to approximately 900 nm^2 and produces a gate-source capacitance, C_{GS}, of 10.8 aF. Nevertheless, the effect of source resistance has more impact on the transistor performance, and as a result transistors with guard rings are used throughout the SRAM design despite larger layout area. Besides the guard ring, another element that increases the SNT layout is the gate metal thickness. Minimum metal thickness has to be approximately 5 nm to form with moderate grain formation and continuous film coverage [1].

Parasitic device and wire capacitances are calculated using ANSOFT's two-dimensional electrostatic solver. Because of the 2 nm apart concentric cylindrical surfaces between gate and source terminals, and the resultant C_{GS} of 10.8 aF, the effective input capacitance of a single transistor has increased from 4.3 aF (the gate oxide capacitance) to 15.1 aF. The gate-drain capacitance, C_{GD}, has stayed the same at 1.7 aF in both layout topologies. The metal-metal interconnect coupling capacitance dictates the total parasitic wire capacitance and is approximately equal to 0.05 aF/nm.

6.4.2 Read and Write Access Times

The 16 × 16 SRAM block is designed to operate between two extreme ambient conditions. The best-case condition is defined so that the transistor ON current reaches its highest value with a 20 % less threshold voltage, 20 % more supply

Table 6.2 The best, the nominal, and the worst ambient conditions

Temperature (°C)	Supply voltage (V)	NMOS V_T (mV)	PMOS V_T (mV)
0	1.2	213	225
27	1.0	266	281
125	0.8	319	337

voltage, and 0 °C operating temperature. Similarly, the worst-case condition causes the ON current to reach its lowest value with a 20 % more threshold voltage, 20 % less supply voltage, and 125 °C operating temperature. Table 6.2 tabulates the values of the best, nominal, and worst ambient temperatures as well as the threshold and the supply voltages.

Propagation delays through address and data paths vary more than 300 %, i.e. the propagation delay at the Address node decreases from 53 ps to 15 ps when the ambient condition is changed from the worst-case to the best-case condition. Similarly, the origin and duration of each control pulse change between 100 % and 600 % with respect to two extreme ambient conditions. For example, the precharge pulse shrinks from 22 ps to 10 ps, EnableWL from 85 ps to 18 ps, SensePulse from 134 ps to 23 ps, Strobe from 134 ps to 20 ps and WritePulse from 78 ps to 28 ps between the worst-case and the best ambient conditions. Therefore, two basic timing rules should always be checked prior to a read or a write operation: (a) an associated control pulse must always follow the data for validation; this rule must especially be verified during the worse-case read and the worst-case write operations, and (b) the precharge pulse must never overlap any data validation pulse.

The longest propagation delays through the address decoder, write and read circuits, and the duration of the control pulses including precharge, EnableWL, SensePulse, Strobe, and WritePulse are shown in Table 6.3a, b for the best and the worst-case conditions, respectively. Each control pulse follows the timing rules outlined above. The worst-case EnableWL is generated approximately 4 ns after the Address signal. The worst-case WritePulse is generated 4 ns after the termination of precharge pulse and terminated 5 ns after the data is written into the 6 T cell. Strobe is produced when the differential potential reaches 50 mV between the Bit and the Bitbar lines. Each control pulse is also well formed and free of superfluous spikes at the rising and falling edges. For example, the best-case WL pulse produced by the combination of the Address signal and EnableWL has 7.6 ps rise time, 6.3 ps fall time, and 15 ps pulse width. Even though its duration almost equals to the sum of rise and fall times, the pulse is well formed for accessing the data in a 6 T cell and its rail-to-rail transition is smooth in both rising and falling edges. The Strobe pulse produced under the best-case ambient conditions is another short pulse with 20 ps duration, 11.6 ps rise time, and 13.2 ps fall time. Nevertheless, it exhibits the same mature pulse characteristics as the WL signal.

Table 6.3 Best and worst-case propagation delay and validation pulse characteristics

	Prop delay (ps)	Pulse origin (ps)	Pulse duration (ps)
(a) The best-case propagation and validation pulse characteristics			
Address	15	–	–
Dataln	5	–	–
Precharge	–	4	10
Enable WL	–	39	18
SensePls	–	71	23
Strobe	–	71	20
WritePls	–	26	28
(b) The worst-case propagation and validation pulse characteristics			
Address	53	–	–
Dataln	21	–	–
Precharge	–	18	22
Enable WL	–	57	85
SensePls	–	92	134
Strobe	–	92	134
WritePls	–	44	78

Table 6.4 Write and read access times, average dynamic power dissipations

	Read access time (ps)	Write access time (ps)	P_{AVER} (read) (μW/col)	P_{AVER} (write) (μW/col)
Best case	78	49	20.7	15.7
Nominal	90	62	11.4	8.2
Worst case	133	98	6.9	5.0

6.4.3 *Power Dissipation*

Average power dissipation is measured as a function of the ambient condition during the read and the write cycles. The power dissipation measurements are taken at 500 MHz and with an output capacitance of 45 aF which corresponds to a fan-out of 3 transistors, each with an effective input capacitance of approximately 15 aF. Table 6.4 summarizes the average dynamic power dissipation per SRAM column at the best-case, nominal-case and worst-case ambient conditions for the read and the write operations. According to this table, a 16 × 16 SRAM dissipates maximum of 331.2 μW during read and 251.2 μW during write operations at 500 MHz. Since the transistor ON current is highest during the best ambient condition and it charges a fixed nodal capacitance, the average dynamic power dissipation is highest during the best-case ambient condition than any other condition at 500 MHz.

6.4.4 SRAM Layout

The 16 × 16 SRAM block is composed of 16 columns of SRAM core, five self-timed circuits, and an address decoder as shown in Fig. 6.8. The total layout area of the block is approximately equal to 3.79 μm by 4.21 μm, which is merely 28 times larger than a 6 T cell in a 65 nm technology [2]. The SRAM core primarily occupies most of the layout area and is equal to 8.25 μm^2. The address decoder and self-timed circuits occupy 1.70 μm^2 and 1.97 μm^2, respectively.

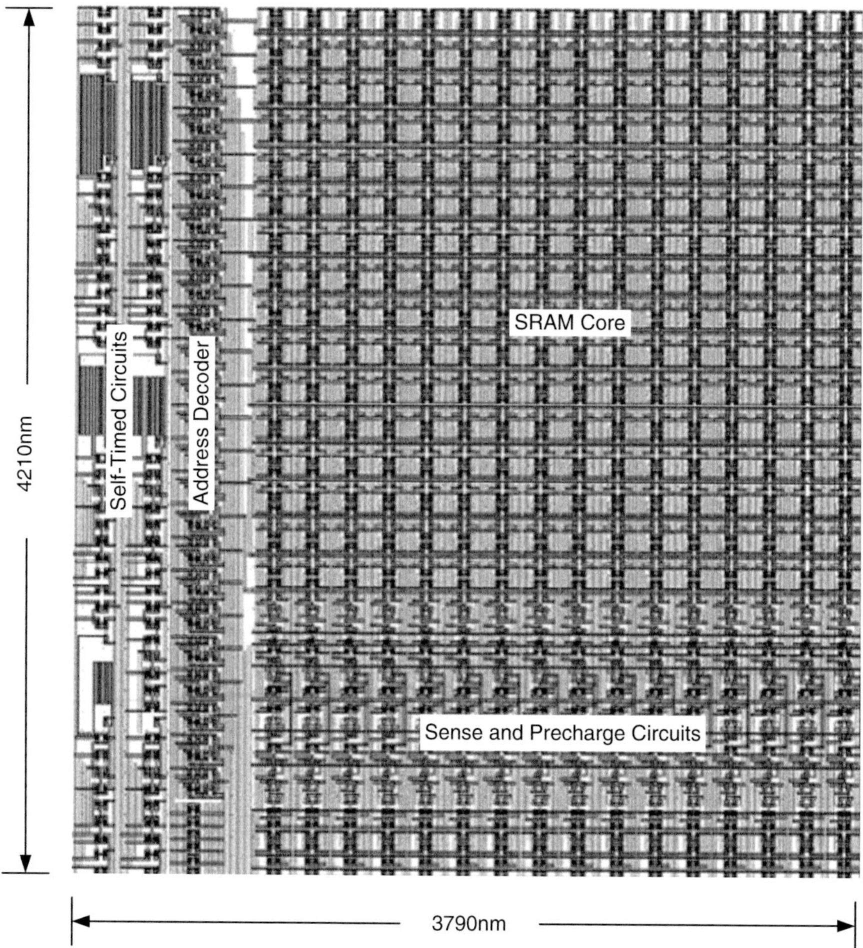

Fig. 6.8 The layout of the 16 × 16 SRAM block. SRAM core, address decoder, and self-timed circuits are shown on the layout

6.5 Summary

A 16 × 16 SRAM block is designed using silicon nanowire technology. In the first section of this chapter, an SRAM architecture including the SRAM core, the read and write circuits, the address decoder, and the self-timed circuits is described. The write and the read operation sequences are explained; waveforms illustrating typical data propagation and validation are presented for each case. In the second section, the performance and dynamic power dissipation figures are given. For example, the worst-case write and read access times for this SRAM block are 98 ps and 133 ps, respectively; the dynamic power dissipation is 20.7 μW per column during a read and 15.7 μW per column during a write operation at 500 MHz. The SRAM layout occupies approximately 16 μm^2.

References

1. Becker JS, Gordon RG (2003) Diffusion barrier properties of tungsten nitride films grown by atomic layer deposition from bis(tert-butylimido)bis(dimethylamido) tungsten and ammonia. Appl Phys Lett 82(14):2239–2241
2. Bai P et al (2004) A 65nm logic technology featuring 35nm gate lengths, enhanced strain, 8 Cu interconnect layers, low-k ILD and 0.57 μm2 SRAM cell. IEDM. pp 657–660

Chapter 7
Field-Programmable-Gate-Array (FPGA)

7.1 Introduction

To be able to implement large-scale SOC designs, minimizing overall power dissipation is a critical [1]. The primary objective of this chapter is to present the results of silicon nanowire technology in a widely utilized prototyping platform called Field-Programmable Gate Array (FPGA). The proposed FPGA architecture in this chapter uses cluster blocks, each of which includes several Look-Up-Tables (LUT) to configure any logic functionality. Each LUT can be configured as a combinatorial logic block or part of a state machine. This flexible configuration is achieved by scan chains implemented inside the cluster block to define the interconnectivity between LUTs and to determine the logic functionality for each LUT. After describing the architectural aspects of the LUT and the cluster, circuit simulation were performed using BSIMSOI SNT models. The chapter reports the results worst-case propagation delays and power dissipation figures of various FPGA circuits and shows typical LUT and the cluster layouts.

7.2 Brief Description of Transistor Design and Modeling

Chapters 1, 2 and 3 show the 3D structure of the device and the complete design process for undoped SNTs. The device design goals were set to achieve minimum static and dynamic power dissipations and minimum intrinsic transient times as discussed in Chapter 3. This criterion resulted in a 2 nm diameter and 10 nm channel length SNTs and the generation of extrinsic BSIMSOI device models to be used for all the circuit simulations in this chapter.

A. Bindal, S. Hamedi-Hagh, *Silicon Nanowire Transistors*,
DOI 10.1007/978-3-319-27177-4_7

7.3 FPGA Architecture

7.3.1 Cluster Architecture

FPGA can be considered as one of the most important application platforms for the SNT technology to test its true potential. An FPGA consists of configuration logic blocks called clusters which contain several LUTs [2]. To program each LUT, scan chains are used. Scan chains are basic shift registers that transmit serial data to configure the logic functionality for each LUT. A typical block diagram of a cluster is shown in Fig. 7.1.

In this figure, an 8-bit wide intercluster bus, line0 through line7, sustains continuous data exchange among clusters. The intercluster data is routed in any available direction using switch boxes placed at four corners of each cluster.

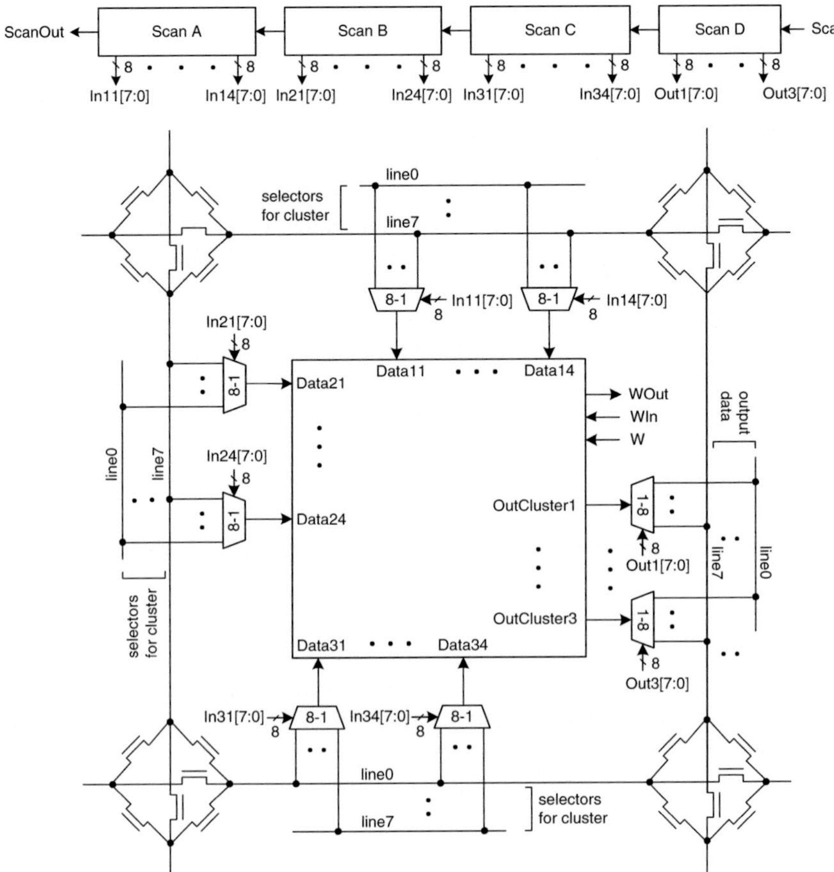

Fig. 7.1 The proposed FPGA architecture with global scan chains

Each bit of the 8-bit intercluster bus is connected to a different switch box composed of six SNTs programmed to guide the incoming data, establishing a configurable path between clusters. A total of 12 data inputs ports, Data11 through Data34, accept incoming signals from neighboring clusters; each input port is connected to the 8-bit wide intercluster bus using an 8-1 MUX. Each cluster has three output ports, OutCluster1, OutCluster2, and OutCluster3, which transfer the cluster data to other clusters for data processing. A particular cluster output can be routed to any one of the eight bits of the intercluster bus using 1-8 DEMUX blocks.

There are two types of scan chains embedded in each cluster. The outputs of the scan chains A, B and C are used as the selector inputs for each 8-1 MUX connected to input ports of the cluster. Once the 8-1 MUX selector is programmed by a scan chain, then incoming signals from neighboring clusters are routed to a specific cluster input for processing. Similarly, scan chain D outputs are used as selectors for each 1-8 DEMUX at each cluster output. Once the selector for each 1-8 DEMUX is programmed, the processed data in a particular cluster is distributed among neighboring clusters for further processing. The serial ports, WIn and WOut, are used to program each LUT of a particular cluster. Serial ports that belong to neighboring clusters are cascaded such that WOut of one cluster is connected to WIn of the other to propagate programming data to all clusters without needing more than one wiring channel. W is another input port that programs each LUT, cluster selector tree and bypass path in each cluster.

Each cluster in Fig. 7.1 is composed of three 4-input Look-Up-Tables (4-LUT) as shown in Fig. 7.2. Each 4-LUT receives external input data, Data11 through Data34, through twelve 4-1 MUXes and can be programmed to implement a primitive or complex logic function requiring up to four inputs. The number of 4-LUTs in each cluster and the intercluster bus width are determined according to the study in [3] to maximize 4-LUT usage and increase available wire utilization between clusters.

There is an additional internal scan chain in each cluster as shown in Fig. 7.2. The primary purpose of this chain is to program all 24 selector inputs, s111 through s342, to control the flow of data into the cluster and to generate program inputs for the three bypass paths in each LUT, bypass1 through bypass3.

7.3.2 4-Input Look-Up-Table (4-LUT)

Each 4-LUT is composed of a 16-bit scan chain at the input and a 16-1 pass-gate MUX tree as shown in Fig. 7.3.

The 16-bit scan chain is used to program a specific logic function for each 4-LUT in the cluster. Each 4-LUT has programming input and output ports, PIn and POut, respectively. PIn of the first LUT, PIn1, receives the programming data directly from WIn of the cluster. POut of the first 4-LUT, POut1, is connected to PIn2 of the second 4-LUT to program the second 4-LUT. Similarly, POut2 of the second 4-LUT is connected to PIn3 to program the third 4-LUT. The serial data at

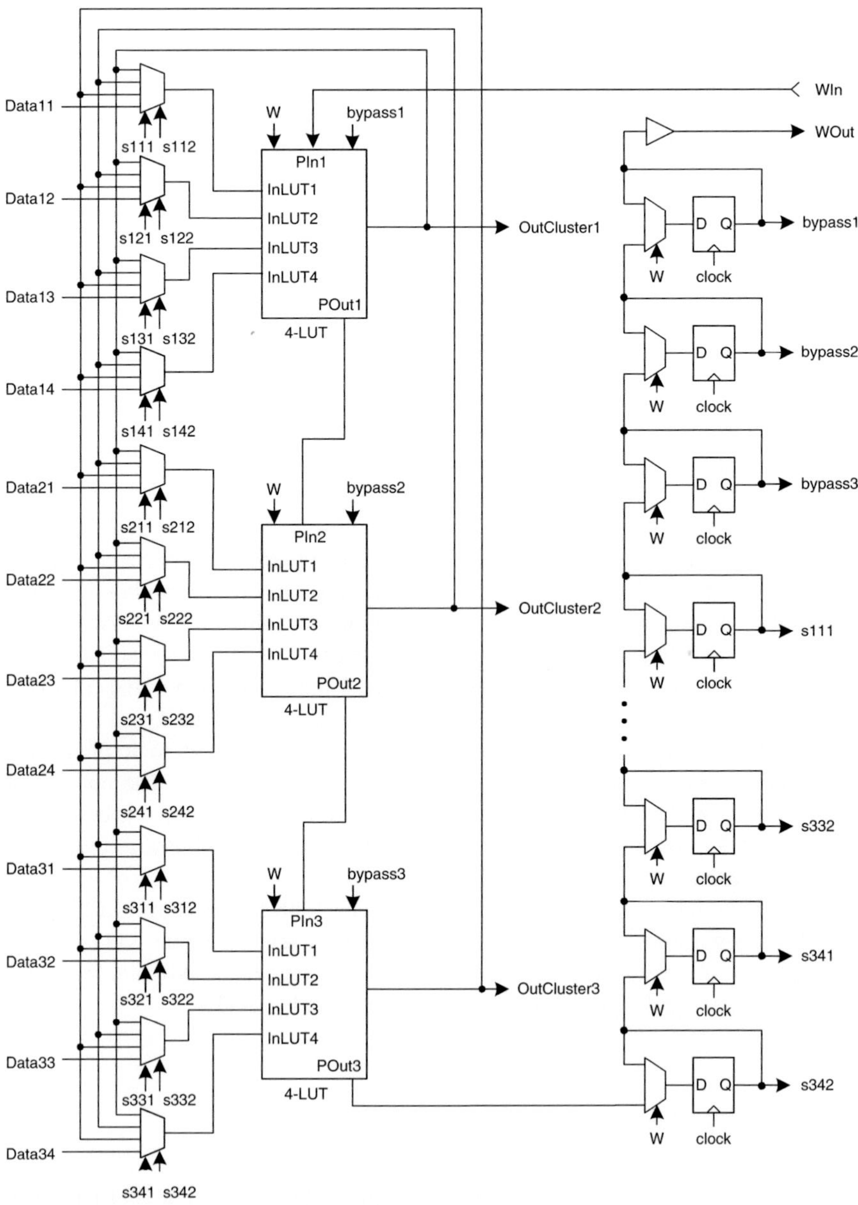

Fig. 7.2 The cluster architecture containing three 4-LUT

the PIn input propagates through the scan chain at the positive edge of clock while the global write signal, W, is kept at logic 1. When scan is finished to program the logic function for each 4-LUT, W is lowered and kept at logic 0 during normal FPGA operation.

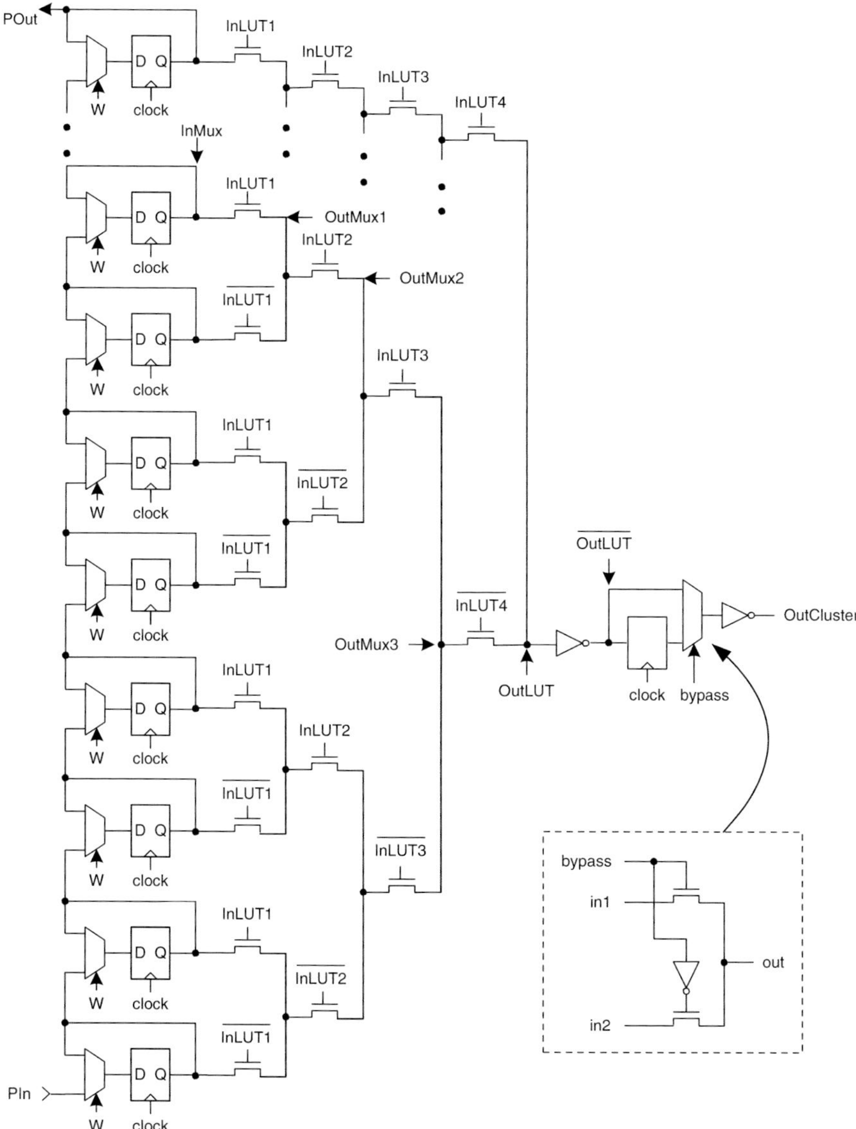

Fig. 7.3 The 4-LUT circuit containing 1 × 16 Look-Up-Table and 16-1 pass-gate MUX

Each selector input to the 16-1 MUX, InLUT1 through InLUT4, is formed either by the cluster inputs routed to this specific 4-LUT (e.g., Data11 through Data14 if referred to the first 4-LUT) or from the outputs of each 4-LUT in the cluster, OutCluster1 through OutCluster3. The output of 16-1 MUX is either registered or routed directly to the output of the 4-LUT via a bypass path.

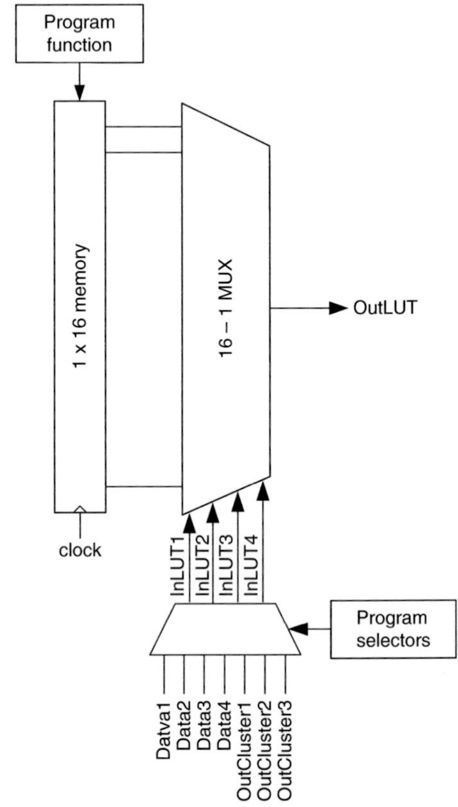

Fig. 7.4 Functional representation of the 4-LUT

The registered 4-LUT outputs are either routed to the inputs of neighboring clusters to form pipelined structures or fed back to the 4-LUT inputs within the same cluster to implement state machines. This flexibility of logic configuration creates an environment to implement any hybrid logic block composed of combinatorial and sequential logic functions.

The functional representation of each 4-LUT is shown in Fig. 7.4. In this figure, 1×16 memory block represents 16-bit input scan chain in Fig. 7.3 responsible for storing a particular logic function. An output is generated for each 4-LUT using a 16-1 pass-gate MUX.

7.3.3 An Example: A 3-bit Carry-Ripple Adder (CRA)

This simple 3-bit Carry-Ripple Adder (CRA) example demonstrates the capability of the proposed FPGA architecture to implement a combinatorial block. A total of six 4-LUTs in two separate clusters are used to implement the 3-bit CRA. The truth

Fig. 7.5 Truth table of a full adder using 4-LUT

A	B	Cin	Data4	Sum	Cout
0	0	0	0 X	0	0
0	0	0	1	0	0
0	0	1	0	1	0
0	0	1	1	1	0
0	1	0	0	1	0
0	1	0	1	1	0
0	1	1	0	0	1
0	1	1	1	0	1
1	0	0	0	1	0
1	0	0	1	1	0
1	0	1	0	0	1
1	0	1	1	0	1
1	1	0	0	0	1
1	1	0	1	0	1
1	1	1	0	1	1
1	1	1	1	1	1

table to generate sum (Sum) and carry-out (Cout) bits of each full adder (FA) in the CRA is shown in Fig. 7.5.

Since each full adder in the CRA has the A, B and carry-in (Cin) inputs, the fourth input to the full adder receives a "don't care" entry, which makes each Sum and Cout output repeat twice in the truth table. The implementation of the 3-bit CRA is shown in Fig. 7.6. The Sum0 output corresponding to the least significant bit of the full adder is generated by storing the Sum column of the truth table in Fig. 7.5 into the 1×16 memory array of the 4-LUT in Fig. 7.6, and programming the cluster scan chain to produce the selector inputs, s111 = s112 = s121 = s122 = s131 = s132 = s141 = s142 = 0. This allows the inputs, Data11 through Data14, to be routed directly to the selector inputs, InLUT1 through InLUT4, of the first 4-LUT, respectively. Therefore, selector inputs for 4-LUT become InLUT1 = Data11 = A0, InLUT2 = Data12 = B0, and InLUT3 = Data13 = Cin0. InLUT4 = Data14 takes a "don't care" input, which converts the 16-1MUX in Fig. 7.3 into two identical 8-1 MUXes. Only one 8-1 MUX output is connected to the OutLUT node in Fig. 7.3 because the InLUT4 input has a "don't care" value. The other 8-1 MUX output stays unconnected. The same approach applies to generate the Cout0 output corresponding to the least significant full adder. The Cout column in Fig. 7.5 is stored in the 1×16 memory array of the second 4-LUT in the cluster. The selector inputs, s211 through s242, are programmed to be equal to 0 such that InLUT1 = Data21 = A0, InLUT2 = Data22 = B0, and InLUT3 = Data23 = Cin0. InLUT4 = Data24 again takes a "don't care" input. The Cout0 output is directly routed to the third 4-LUT in the same cluster and also to the first 4-LUT of the second cluster to generate the Sum1 and Cout1 outputs of the next significant full adder bit of the 3-bit CRA.

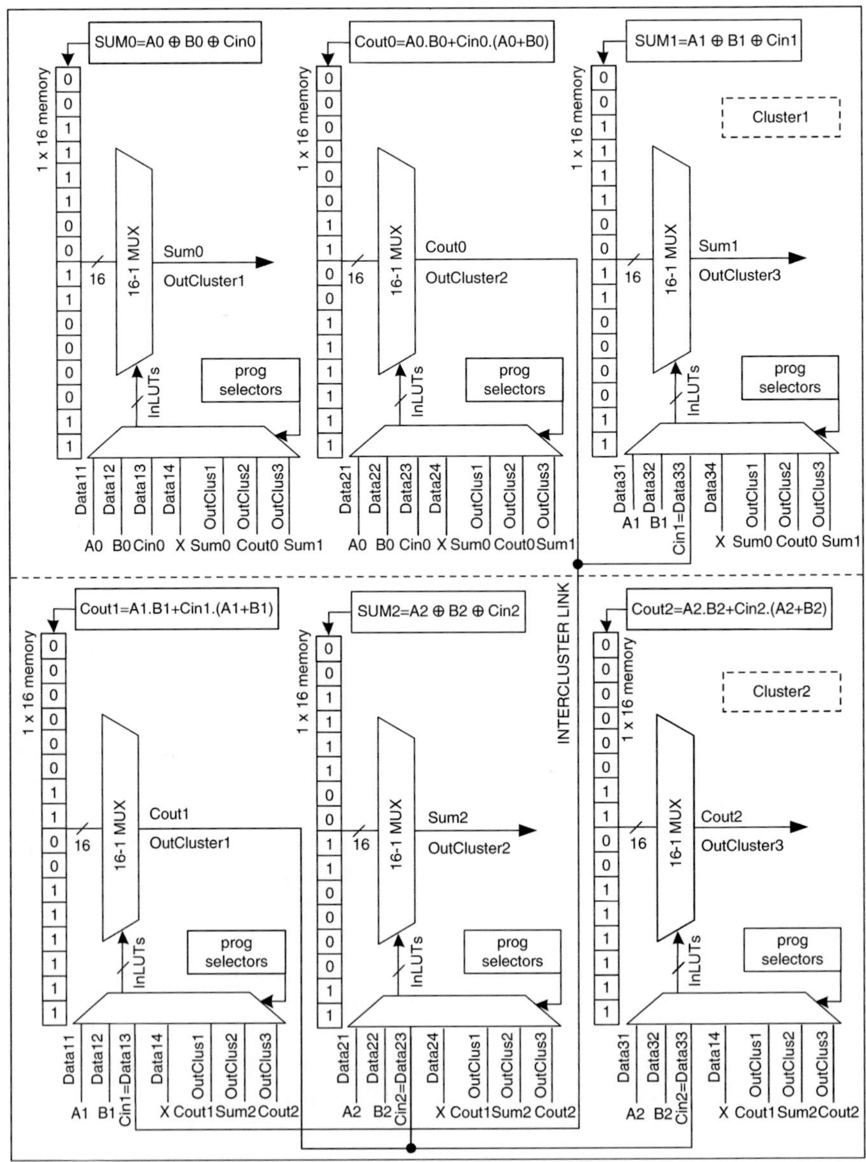

Fig. 7.6 3-bit Carry-Ripple Adder circuit using two separate clusters

Similarly, the Cout1 output is routed to the inputs of the second and third 4-LUTs of the second cluster to generate the Sum2 and Cout2 outputs, respectively. The Bypass1, Bypass2, and Bypass3 inputs are programmed to be at logic 1 to make all three bits of the CRA purely combinatorial and unregistered.

7.4 FPGA Circuit Characteristics

7.4.1 4-LUT Worst-Case Propagation Delays

Figure 7.7 shows the worst-case propagation delays at the internal nodes of the 4-LUT. In this figure, all uncomplemented InLUT inputs, InLUT1 through InLUT4, are assumed to be connected to logic 1. When the signal reaches the internal node, OutMUX1, the logic 1 level at that node drops by an amount equal to NMOS threshold voltage, V_{TN}, as expected. This trend continues at the subsequent nodes from OutMUX2 to OutLUT until normal logic levels are restored at the OutLUT node by an inverter. The OutCluster node shows a slower rise time due to a 100 aF load capacitor. Typical worst-case delays in Fig. 7.7 are 4.7 ps from the clock input to the InMUX node, 8.9 ps from the clock input to the OutLUT node, and 20.2 ps from the clock input to the OutCluster node.

Figure 7.8 shows the worst-case delay from the clock input to the OutCluster node as a function of load capacitor and it is expressed as $T_{CLK\text{-}OUTCLUSTER} = 10.8 + 0.09$ CLOAD in ps.

7.4.2 Intercluster Propagation Delays

The intercluster delays are explained in Fig. 7.9. In this figure, a worst-case signal path is shown when two clusters are placed in a diagonal fashion. A typical intercluster signal path starts from the OutCluster port of the source cluster and travels through the pass-gate transistor in the 1-8 DEMUX stage, an intercluster wire along the *y*-axis of the source cluster, a pass-gate transistor at the first switch box, the second intercluster wire along the *y*-axis of the destination cluster, two pass-gate transistors in the second switch box, the third intercluster wire half the

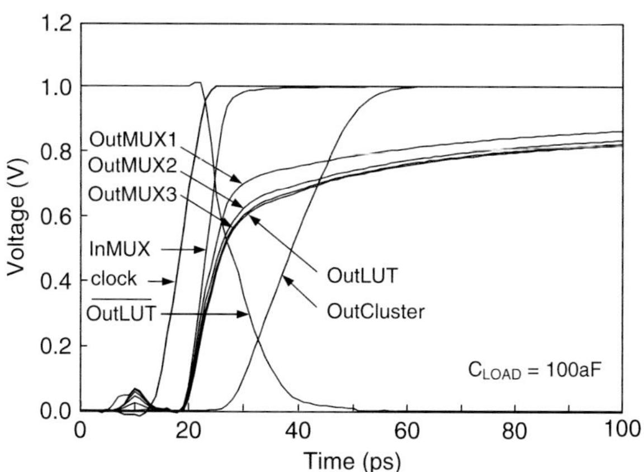

Fig. 7.7 Post-layout, worst-case propagation delays through 4-LUT

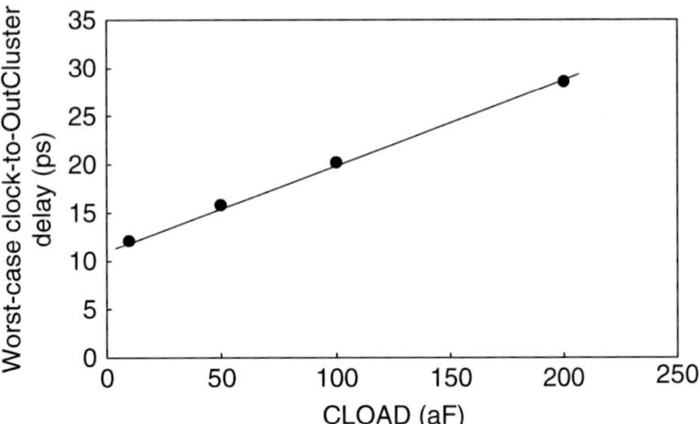

Fig. 7.8 Post-layout, worst-case propagation delay from clock input to 4-LUT output as a function of output load

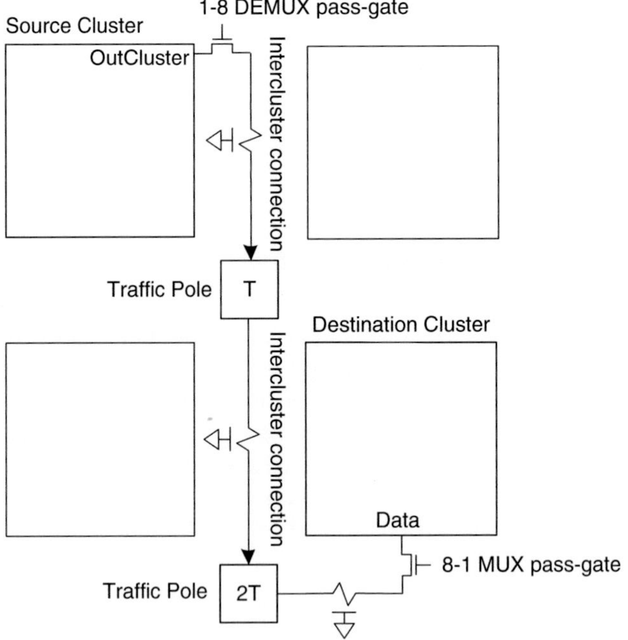

Fig. 7.9 Intercluster wiring topology

length along the *x*-axis of destination cluster, and a pass-gate transistor in the 8-1 MUX stage before arriving at the data input of the destination cluster.

Typical worst-case delay waveforms at the OutCluster port of the source cluster, the Data port, and the InLUT1 nodes of the destination cluster are shown in Fig. 7.10. In this figure, fan-out is equal to one because only one destination cluster is connected to the source cluster. The worst-case intercluster delay as a function of fan-out is shown in Fig. 7.11 where fan-out is defined as the number of destination clusters. In this figure, the delay analysis is performed on a diagonally placed source and destination clusters two and four cluster lengths apart.

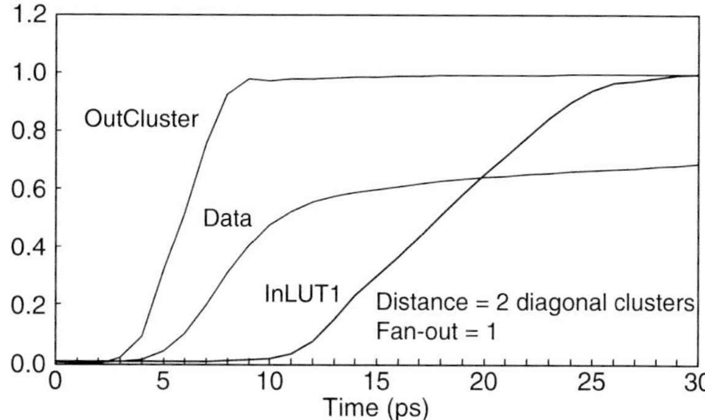

Fig. 7.10 Post-layout, worst-case intercluster wiring delay between two neighboring clusters diagonally placed two clusters apart

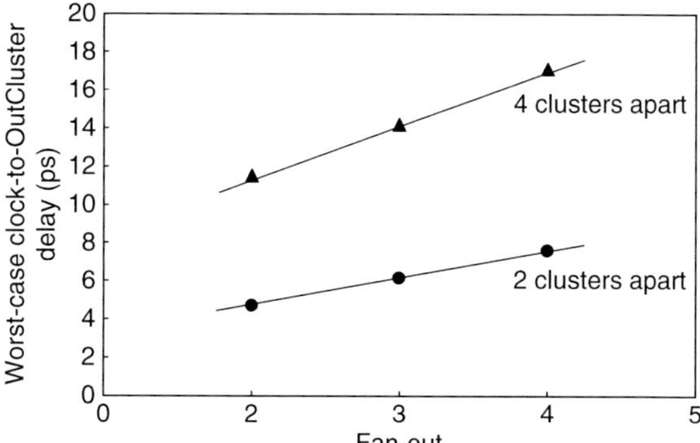

Fig. 7.11 Post-layout, worst-case intercluster wiring delay between two diagonally placed neighboring clusters as a function of fan-out (the number of destination clusters)

7.4.3 4-LUT Power Dissipation

Since 16-1 input MUX of the 4-LUT is configured using only pass-gate transistors, it is quite possible that the InLUT (InLUT1 through InLUT4) and the $\overline{\text{InLUT}}$ ($\overline{\text{InLUT1}}$ through $\overline{\text{InLUT4}}$) inputs may not arrive at the same time and the overlap between these signals may develop varying dynamic power dissipations for 4-LUT. Figure 7.12 shows the worst-case dynamic power dissipation as a function of signal overlap for 2 GHz, 4 GHz, and 10 GHz frequencies. To obtain the 2 GHz data, a 200 ps pulse with a 500 ps period is applied to all InLUT and $\overline{\text{InLUT}}$ inputs. However, the pulse applied to InLUT inputs is forced to overlap with the pulse applied to $\overline{\text{InLUT}}$ inputs by 20 ps, 40 ps, 60 ps, 80 ps and 100 ps intervals. Subsequently, total dynamic power dissipation is measured and averaged within 500 ps window for each case. The same procedure is applied to obtain the 4 GHz and 10 GHz data. However, the pulse width and period are both decreased to 100 ps and 250 ps for 4 GHz, and to 50 ps and 100 ps for 10 GHz, respectively. Both the level and the slope of power dissipation are observed to increase with increasing frequency. However, when there is no signal overlap, the power dissipation in 4-LUT (1 μW at 2 GHz and 2.8 μW at 10 GHz) is found to be independent of any linear frequency-power relationship.

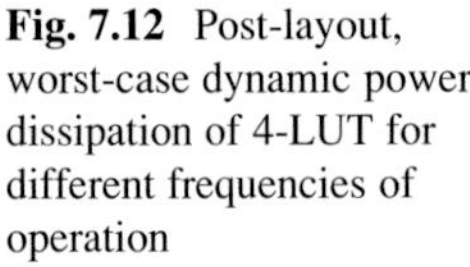

Fig. 7.12 Post-layout, worst-case dynamic power dissipation of 4-LUT for different frequencies of operation

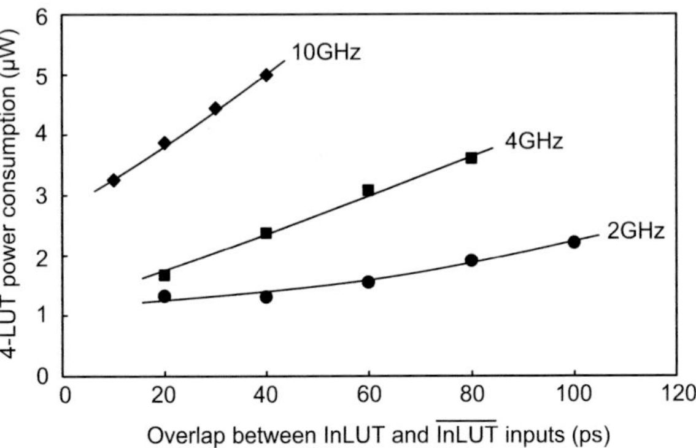

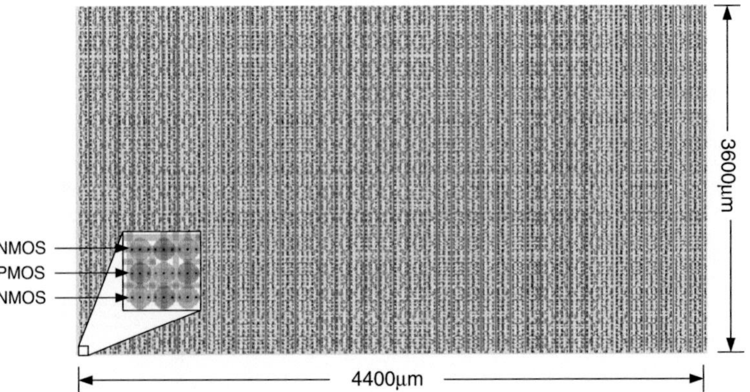

Fig. 7.13 The cluster layout

7.4.4 Flip-Flop Characteristics

Each cluster contains 93 D-type flip-flops, which are used for both scanning control inputs and logic configuration. A typical flip-flop used in this design has 5 ps set-up time, 1 ps hold time, and 11.4 ps clock-to-output delay at a load capacitor of 10 aF. Its worst-case power dissipation is 1.2 μW at 10 GHz.

7.4.5 Cluster Layout

Figure 7.13 shows a cluster layout that contains three 4-LUTs. Note that this layout contains alternating NMOS-PMOS placement methodology in a fabric-like matrix as shown in the subpicture of Fig. 7.13. Every NMOS transistor is surrounded by 4 PMOS transistors and vice versa in a cross-bar form. This way, fabrication-related issues such as uniform vertical crystal growth of transistor bodies, Reactive Ion Etch (RIE) lag etc. may be minimized. Each 4-LUT input is placed at a different cluster boundary: the first 4-LUT inputs are placed at the top, the second at the left, and the third at the bottom. The cluster outputs and control signals are placed at the right boundary of the cluster.

7.5 Summary

This study investigates the outcome of using silicon nanowire transistor circuits in FPGAs containing scan chains. Each FPGA cluster in this architecture is composed of three 4-LUTs. Each cluster receives input data from other clusters through its twelve data inputs and sends data to neighboring clusters from its three outputs. Each 4-LUT in the cluster can be programmed to implement a specific logic function with a maximum of four inputs or can be configured as part of a state machine. Scan chains are used to program the data path in the cluster, to determine the logic function for each 4-LUT and most importantly to minimize wiring channels in the entire FPGA array. Intercluster communication is established by an 8-bit wide bus architecture interconnected with switch boxes that contain six transistors and are placed at the corners of each cluster. Post-layout, worst-case propagation delay for a 4-LUT is 20.8 ps from the rising edge of the clock signal to its output. The worst-case intercluster wire delay is 4.8 ps between two diagonally placed adjacent clusters and increases to 11.2 ps between two neighboring clusters diagonally placed at four cluster lengths away. The average worst-case dynamic power dissipation of a 4-LUT is 1 μW at 2 GHz and 2.8 μW at 10 GHz if there is no overlap between uncomplemented and complemented selector pulses for the 4-LUT. If there is an overlap between the selector pulses, power dissipation increases by 12.5 nW/ns of the overlap at 2 GHz and 60 nW/ns of the overlap at 10 GHz. Cluster layout contains vertical silicon nanowire transistors placed in a fabric matrix where each NMOS (PMOS) transistor has four neighboring PMOS (NMOS) transistors.

References

1. Choudhary P, Marculescu D (2009) Power management of voltage/frequency island-based systems using hardware-based methods. IEEE Trans VLSI Syst 17(3):427–438
2. Bindal A, Hamedi-Hagh S, Ogura T (2008) Silicon nanowire technology for applications in the field programmable gate array architectures. J Nano Opto 3:113–122
3. Ahmed E, Rose J (2004) The effect of LUT and cluster size on deep-submicron FPGA performance and density. IEEE Trans VLSI Syst 12(3):288–298

Chapter 8
Integrate-and-Fire Spiking (IFS) Neuron

8.1 Introduction

This chapter applies the benefits of silicon nanowire technology yet to another interesting digital design area: designing a digital neuron. A neuron is an essential element to mimic brain-like processing functions such as recognition, perception, and it is composed of three parts: dendrites, soma, and axon. Electric impulses are transmitted to dendrites via a few thousand synapses, a postsynaptic potential is generated at soma, and an impulse is generated and distributed among neighboring neurons via synapses attached to the axon. To associate this behavior in VLSI, a realistic implementation of this device is necessary. This chapter uses extrinsic BSMSOI SNT models generated in Chapter 3 to perform circuit design and simulations of a digital neuron. It reports electrical data including performance and power consumption figures of a single neuron cell and its layout.

The early neural models that use Pulse Rate Coding (PRC) were implemented in analog [1] and digital [2] domains using a back-propagation algorithm. Pulse Density Modulation (PDM) was another algorithm that measured the rate of pulsed events [3]. Recently, a more realistic approach to model neurons is the Spiking Response Model (SRM) [4, 5] that directly mimics Hopfield's equation [6]. Various VLSI implementations favor a Leaky Integrate-and-Fire (LIF) SRM for computational accuracy [7–9]. Others modified the SRM with fuzzy logic to create a new generation of neurons for character recognition [10], formed a quantizer neuron and multifunctional layered network for image recognition and learning [11], and mimicked chaotic processes in the brain [12]. The SRM was also reduced to a multi-nanodot floating gate MOSFET form for achieving high density and low power neuron cells in one chip [13].

The primary objective of this study is to design a new, ultra-compact integrate-and-fire spiking (IFS) neuron dissipating substantially lower power and occupying less die area compared to earlier silicon designs [9, 12] using vertical, undoped SNTs grown on SOI substrates.

A. Bindal, S. Hamedi-Hagh, *Silicon Nanowire Transistors*,
DOI 10.1007/978-3-319-27177-4_8

8.2 Brief Description of Transistor Design and Modeling

Both NMOS and PMOS transistors are enhancement type with undoped silicon bodies and metal gates constructed perpendicular to the SOI substrate as shown in Chapter 1 and then resumed in Chapter 3. The optimal body dimensions of 2 nm radius and 10 nm channel length are determined according to the minimum static and dynamic power dissipations and minimum intrinsic transient times for each NMOS and PMOS SNT. Based on these particular transistor dimensions, the BSIMSOI device models were created in Chapter 3, and subsequently used in this chapter for all circuit simulations.

8.3 IFS Neuron

8.3.1 IFS Neuron Firing Principle

Figure 8.1 shows the firing principle of a single IFS neuron. The neuron cell or "soma" is composed of three parts: the "sum" forms a postsynaptic potential, the "threshold" determines the maximum amplitude for the postsynaptic potential to generate a pulse, and the "fire" forms a positive or a negative amplitude pulse according to the amplitude of the postsynaptic potential. Synaptic pulses at the input of a neuron cell can be excitatory or inhibitory, depending on their amplitude. A +1 V input pulse defines the excitatory input while -1 V defines the inhibitory input. In this figure, five 1000 ps wide excitatory synaptic pulses are asynchronously received by dendrites. A postsynaptic potential is formed at the sum node of the neuron. If the peak of the postsynaptic potential exceeds a certain positive threshold, the neuron produces a +1 V high, 1000 ps wide pulse at its output. This pulse is distributed among other neurons through its axon. Similarly, a negative postsynaptic potential may be accumulated. If this potential exceeds a negative threshold, a -1 V high, 1000 ps wide pulse is produced at the output.

8.3.2 IFS Neuron Design

The IFS neuron schematic is shown in Fig. 8.2. The circuit is composed of four sections; each section is bounded by dashed lines in the schematic.

The first section is where all synaptic inputs are connected to a common node called sum where a sum of capacitors form a postsynaptic potential. The reason for using capacitive coupling to form synaptic junctions is that synaptic connections between two biological neuron cells are not directly connected; electrical impulses are passed from one neuron to the next by transmitting and receiving dendrite ends [6]. The cell threshold voltage is determined by capacitive charge-sharing and is simply changed by altering the total capacitance value at the sum node. For this

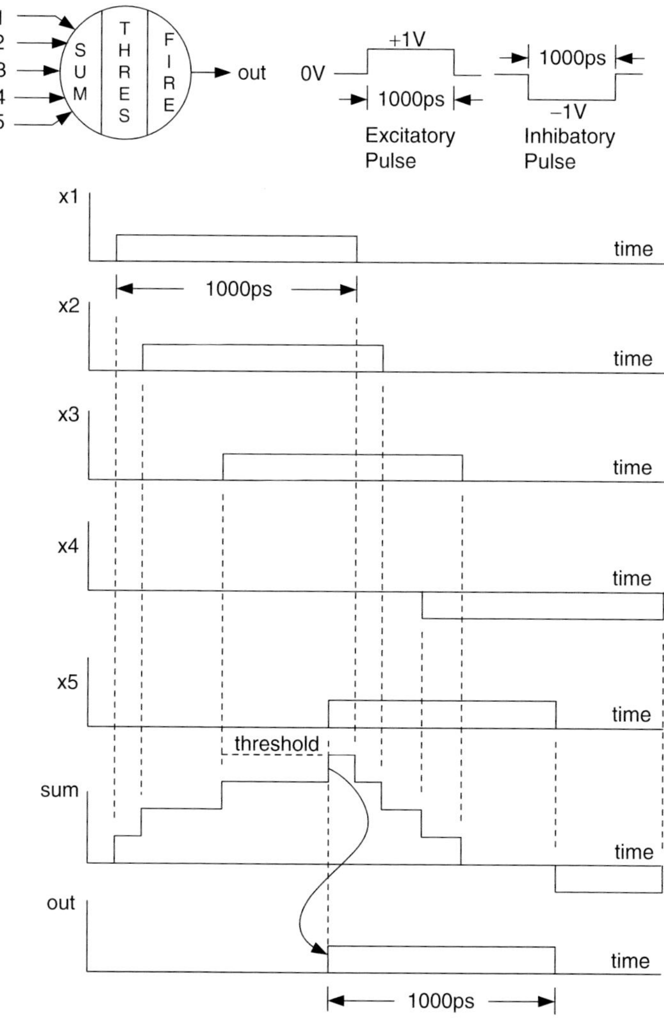

Fig. 8.1 The IFS neuron principle

particular neuron circuit, ±0.8 V is defined as the threshold voltage which generates either +1 V or −1 V output pulse, respectively. This threshold voltage can be changed by adding or subtracting more capacitance at the sum node. Note that Cg corresponds to the gate capacitance of a single NMOS transistor, which is equal to 2.16 aF. Higher value capacitances are formed by connecting the gates of multiple NMOS transistors in parallel.

The second section is a self-time circuit composed of a latch followed by an RC delay element and a simple PMOS charge pump, which produces a +1 V output pulse. The waveforms generated by this section are shown in Fig. 8.3 with contoured arrows, each of which indicates how the next waveform is produced.

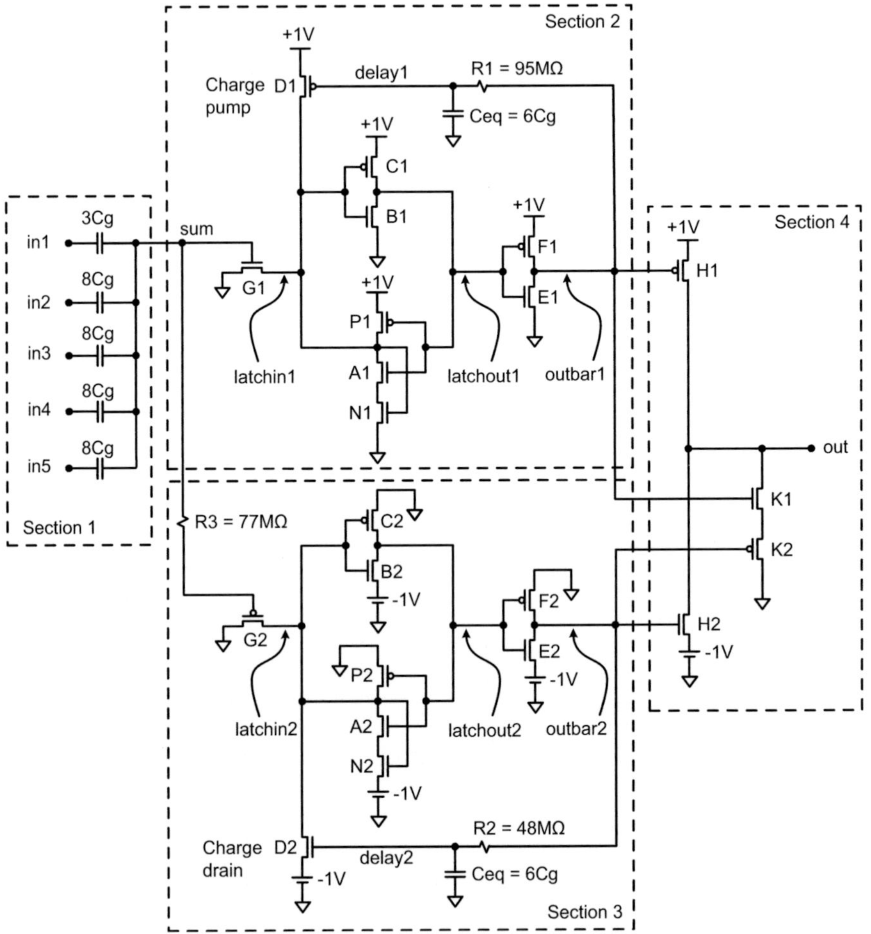

Fig. 8.2 The IFS neuron schematic

When the postsynaptic potential at the sum node exceeds +0.8 V, the NMOS pass gate transistor, G1, turns on and changes the latchin1 node in Fig. 8.2 from +1 V to 0 V.

Since the fighting between the transistors, G1 and P1, is significantly reduced by the addition of a series NMOS transistor, N1, the latchout1 node changes to +1 V, turns on the NMOS transistor, E1, and pulls down the outbar1 node to 0 V after a short delay. Subsequently, the PMOS transistor, H1, at the output stage turns on and charges the output node to +1 V. Following the voltage drop at the outbar1 node, the voltage at the delay1 node also decays towards 0 V with an RC delay produced by R1 = 95 MΩ and Ceq = 6Cg = 13.2 aF. This decay eventually turns on the PMOS transistor, D1, which starts pumping charge to the latchin1 node. Consequently, this node is pulled up to +1 V which triggers the latch and changes the outbar1 node back to +1 V. This rise at outbar1 turns off the transistor, H1, turns on the transistor,

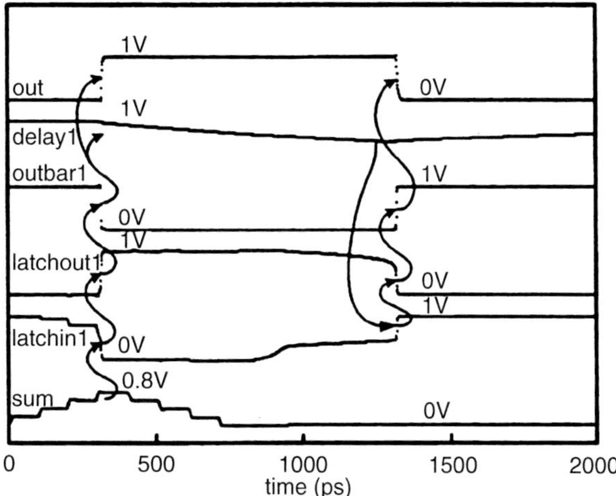

Fig. 8.3 Simulation waveforms of the IFS neuron generating an excitatory pulse whose amplitude is +1 V

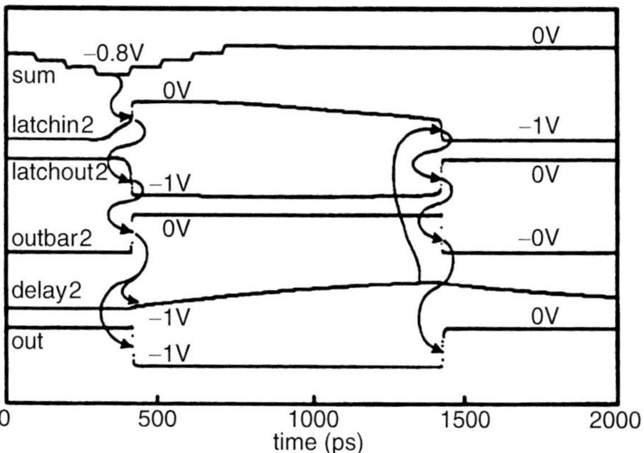

Fig. 8.4 Simulation waveforms of the IFS neuron generating an inhibitory pulse whose amplitude is –1 V

K1, and pulls down the output to 0 V. The width of the resultant output pulse in Fig. 8.3 is measured approximately to be 1006 ps.

The third section represents another self-time circuit composed of a latch, an RC element, and an NMOS charge drainage device, which generates a –1 V output pulse. The waveforms generated by this section are shown in Fig. 8.4. Similar to the self-time circuit operation explained above, when postsynaptic potential at the sum node exceeds –0.8 V, the PMOS pass gate transistor, G2, turns on and changes the latchin2 node from –1 V to 0 V. This voltage triggers the latch, changes the latchout2 node to –1 V and the outbar2 node to 0 V after a short delay. The NMOS transistor, H2, at the output circuit turns on and the output is pulled down to –1 V. Following the voltage rise at the outbar2 node, the voltage at the delay2 node also increases towards 0 V with an RC delay specified by R2 = 48 MΩ and Ceq. The rise at the delay2 node turns on the NMOS transistor, D2, which discharges

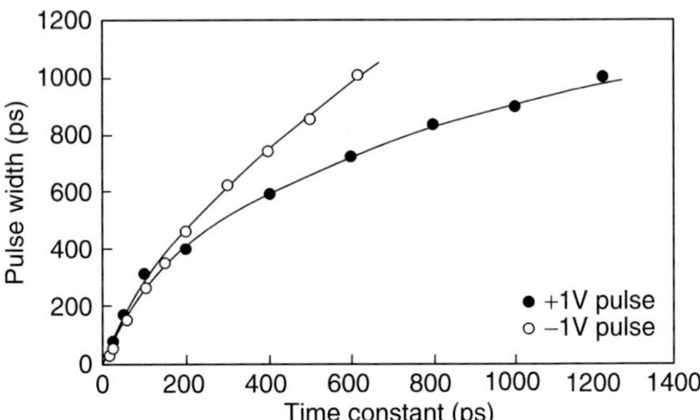

Fig. 8.5 Post-layout pulse width as a function of built-in time constant of the IFS neuron designed with 10 nm effective channel length and 2 nm body radius NMOS and PMOS transistors

the latchin2 node back to −1 V. This voltage, on the other hand, triggers the latch and pulls the outbar2 node to −1 V. Consequently, the NMOS transistor, H2, turns off, the PMOS transistor, K2, turns on, and the output is pulled up to 0 V. The width of the resultant out pulse in Fig. 8.4 is measured approximately to be 1011 ps.

The fourth section is the output buffer stage responsible for maintaining the cell output at 0 V when both H1 and H2 are turned off. However, when the postsynaptic potential at the sum node exceeds 0.8 V, this stage charges the output node to 1 V through H1. When the voltage drops below −0.8 V, the same stage discharges the output node to −1 V through H2.

The neuron cell is designed to operate with excitatory or inhibitory synaptic input pulses that do not decay with time. However, if a conventional LIF neuron [4, 13] needs to be realized, the internal sum node should also include a resistor in parallel with Cg to produce the desired decay in Fig. 8.2.

Figure 8.5 shows the pulse width duration as a function of time constant by adjusting the individual values of resistor and capacitor to control the delay. The capacitor, Ceq, is formed by multiples of 2.16 aF since this value is equal to the gate capacitance of an NMOS transistor. The resistors, R1 and R2, may be formed by a combination of NMOS transistors connected in series. To reduce the number of transistors to five, for example, the gate bias voltage must be reduced to 0.2 V, which requires an additional supply voltage. This particular IFS neuron circuit generates approximately 1000 ps wide output pulses with the resistor and capacitor values shown in Fig. 8.2.

Figure 8.6 shows the error in pulse width as a function of refractory period. The refractory period describes the minimum time interval between consecutive neuron "firings." In other words, after firing a pulse, the neuron cell has to wait for a certain time called the refractory period before it can fire again. Figure 8.6 is formed by forcing the neuron to generate two consecutive 1000 ps wide output pulses and measuring the pulse width error in the secondary pulse as a function of refractory period. For example, 57 % pulse width error to generate a +1 V pulse indicates that the secondary pulse width is reduced more than half the primary pulse width if the refractory period is decreased to 1.1 ns.

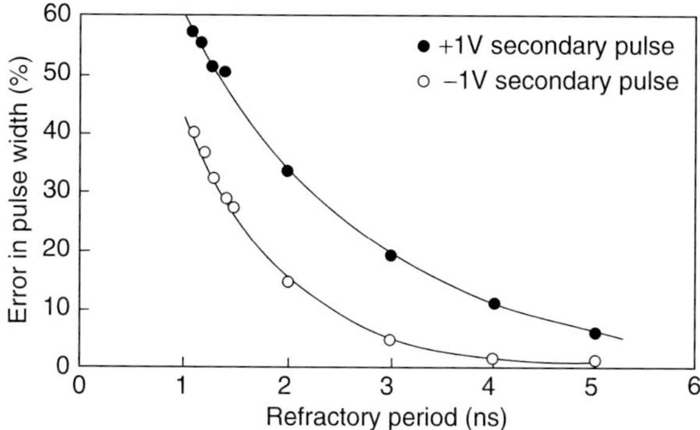

Fig. 8.6 Post-layout error in consecutive pulse widths generated by the IFS neuron as a function of refractory period

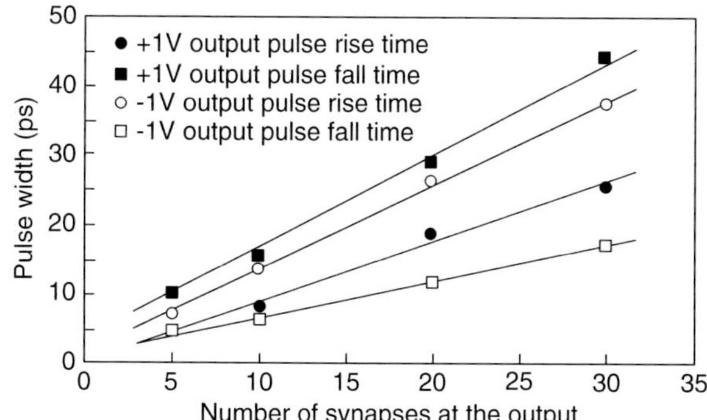

Fig. 8.7 Post-layout worst-case transient time characteristics of the IFS neuron as a function of synaptic load capacitance

8.3.3 *Transient Response and Power Dissipation*

Figure 8.7 shows the worst-case rise and fall times of the output pulse generated by the IFS neuron as a function of synaptic input capacitance, which corresponds to the series combination of 8Cg (17.6 aF) and two pass gate capacitances in parallel. The rise time of +1 V output pulse is equal to 5 ps for a load of five synapses and increases by 0.84 ps per synaptic input connected to the neuron axon. Similarly, the fall time is 10.1 ps for a load of five synapses and increases by 1.3 ps per synapse. The −1 V output pulse reveals comparable number: the rise time of the −1 V output pulse is 7.6 ps for a load of five synapses and increases by 1.2 ps per synapse; the fall time is 5 ps for a load of five synapses and increases by 0.53 ps per synapse.

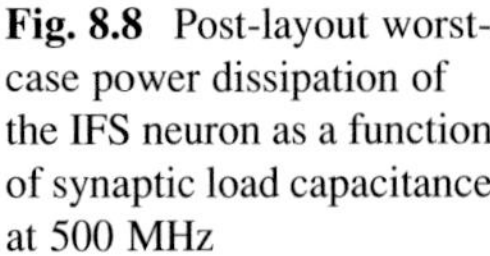

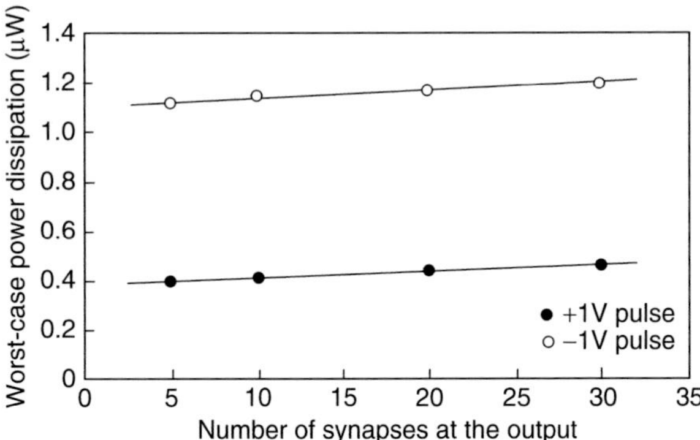

Fig. 8.8 Post-layout worst-case power dissipation of the IFS neuron as a function of synaptic load capacitance at 500 MHz

Figure 8.8 shows the worst-case dynamic power dissipation as a function of synaptic input capacitance. The worst-case power dissipation is obtained by sending different combinations of single 1000 ps wide pulses to all five inputs of the IFS neuron within a 2 ns time interval, measuring for each combination the average value of the power supply current, and finally selecting the combination that draws the maximum average current. The worst-case power dissipation to generate a +1 V pulse is 0.397 μW for an output load of five synapses and increases by 2.5 nW per synapse at 500 MHz. The worst-case power dissipation to generate a −1 V pulse, on the other hand, increases to 1.12 μW for an output load of five synapses and 2.8 nW for each additional synapse.

8.3.4 *IFS Neuron Cell Layout*

Following the design of the IFS neuron layout, measurements were conducted before and after parasitic layout extraction and compared with each other to understand the effects of parasitic wire resistance, capacitance, and contact resistance on circuit performance. The effect of layout parasitics on transient performance is not substantial. The worst-case transient time increases by 8.7 % for a fan-out of one synapse and less than 2.4 % for a fan-out of 30 synapses after parasitic extraction. The worst-case dynamic power dissipation, on the other hand, increases by 9.3 % for a fan-out of one synapse and 3.6 % for a fan-out of 30 synapses after parasitic extraction.

Figure 8.9 shows the layout of the IFS neuron which occupies an area of 550 nm by 500 nm. The transistor names are indicated on the layout to match the schematic given in Fig. 8.2. A recent six-transistor SRAM cell designed in a 65 nm technology occupied a cell area of 0.57 μm^2 [14]. The layout area of the neuron is approximately 0.27 μm^2 which is about two times smaller than the six-transistor SRAM cell and contains 12 times more transistors.

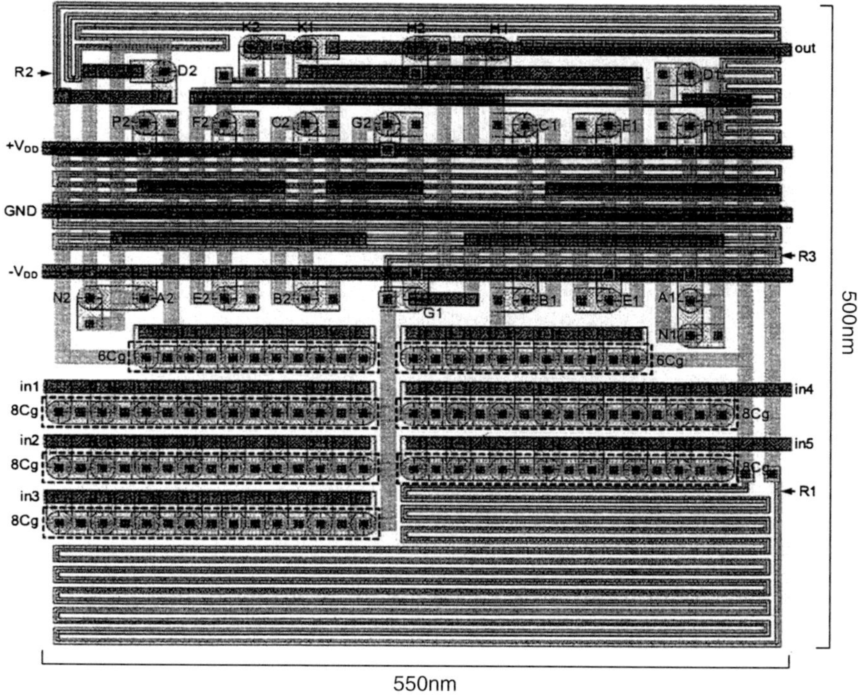

Fig. 8.9 The IFS neuron layout using 10 nm effective channel length and 2 nm body radius NMOS and PMOS nanowire transistors

8.4 Summary

In this chapter, a new integrate-and-fire spiking neuron is designed using dual work function, vertical, undoped silicon nanowire transistors with 10 nm channel length and 2 nm radius. This optimum device geometry has produced NMOS and PMOS transistors with less than 1 pA OFF current and minimal intrinsic energy while generating ON currents in the neighborhood of 10 μA, promoting short intrinsic transient time. The threshold voltage of each transistor was adjusted to 300 mV by fine-tuning gate metal work function of NMOS and PMOS SNT to produce a sufficient noise margin. The IFS neuron circuit is operated with excitatory and inhibitory synaptic input pulses whose amplitudes are changed between 0 to +1 V and 0 to −1 V, respectively. The neuron cell produces excitatory pulses at its output between 47 and 1006 ps and inhibitory pulses between 29 and 1011 ps according to the adjustable RC time constant. The refractory period to generate a 1000 ps wide excitatory or inhibitory pulse is approximately 5 ns; shorter refractory periods gradually decrease the pulse width of consecutive output pulses. The worst-case rise time of a +1 V output pulse is 5 ps at a load capacitance of five synaptic inputs and increases by 0.84 ps per synapse; the worst-case fall time is 10.1 ps and increases by 1.3 ps per synapse. Similarly, the worst-case rise time of a −1 V output pulse is 7.6 ps at a load capacitance of five synaptic inputs and increases by 1.2 ps per synapse; the worst-case fall time is 5 ps and increases by 0.53 ps per synapse. To generate a +1 V output pulse, the IFS neuron consumes 0.397 μW for an output load of five synapses at 500 MHz and its power dissipation increases by 2.5 nW per synapse. Similarly, the generation of a −1 V output pulse requires 1.12 μW power consumption for an output load of five synapses at 500 MHz and increases by 2.5 nW per each additional synapse. Excluding the power dissipation in local and global interconnects, a 1 W chip fabricated with the proposed silicon nanowire technology can accommodate in excess of 800,000 neurons, each of which consumes a maximum of 1.19 μW with 30 output synapses. The neuron cell occupies a layout area of 0.27 μm^2, which is estimated to be 26 times smaller than a conventional layout with the same number of transistors in a 65 nm technology.

References

1. Marie T, Amemiya Y (1994) An all-analog expandable neural network LSI with on-chip backpropagation learning. IEEE J Solid State Circ 29:1086–1093
2. Yasunaga M, Masuda N, Yagyu M, Asai M, Shibata K, Ooyama M, Yamada M, Sakaguchi T, Hashimoto M (1993) A self-learning digital neural network using wafer-scale. LSI IEEE J Solid State Circ 28:106–113
3. Hirai Y, Yasunaga M (1996) A PDM digital neural network system with 1.000 neurons fully interconnected via 1 000 000 6-Bit synapses. Proc. Int. Conf Neural Information Processing. pp 1251–1256
4. Maass W (1999) Pulsed neural networks. MIT Press, Cambridge, MA
5. Maass W (1997) Networks of spiking neurons: the third generation of neural network models. Neural Netw 10:1659–1671
6. Hertz J, Krogh A, Palmer R (1993) Introduction to the theory of neural computation. Addison-Wesley, Reading, MA
7. Fusi S, Annunziato M, Badoni D, Salamon A, Amit D (2000) Spike-driven synaptic plasticity: theory, simulation, VLSI implementation. Neural Comput 12:2227–2258
8. Hafliger P, Mahowald M, Wans L (1997) A spike based learning neuron in analog VLSI advances in neural information. In: Processing systems. MIT Press, Cambridge, MA
9. lndiveri G, Chicca E, Douglas R (2006) A VLSI array of low-power spiking neurons and bistable synapses with spike-timing dependent plasticity. IEEE Trans Neural Networks 17:211–221
10. Yamakawa T (1996) Silicon implementation of a fuzzy neuron. IEEE Trans Fuzzy Syst 4:488–501
11. Maruno S, Kohda T, Nakahira H, Sakiyama S, Mamyama M (1994) Quantizer neuron model and neuroprocessor-named quantizer neuron chip. IEEE J Sel Areas Commun 12:1503–1509
12. Horio Y, Aihara K, Yamamoto O (2003) Neuron-synapse IC chip-set for large-scale chaotic neural networks. lEEE Trans Neural Networks 14:1393–1404
13. Marie T, Matsuura T, Nagata M, Iwata A (2003) A multi-nano-dot floating-gate MOSFET circuit for spiking neuron models. IEEE Trans Nanotechnol 2:158–164
14. Bai Pet (2004) A 65 nrn logic technology featuring 35 nm gate lengths, enhanced strain, 8 Cu interconnect layers SRAM. IEDM. pp 657–660

Chapter 9
Direct Sequence Spread Spectrum (DSSS) Baseband Transmitter

9.1 Introduction

As today's mixed signal chips often require both digital and analog system components on the same substrate, an interesting area of application for silicon nanowire technology is to build a Direct Sequence Spread Spectrum (DSSS) baseband transmitter. A basic baseband transmitter contains an 8-Phase Shift Keying (8-PSK) modulator, a fourth order PN generator, a binary bit mapper, and two bit multipliers. This chapter uses the BSIMSOI models of NMOS and PMOS SNTs developed in Chapter 3 to simulate all transmitter circuits and shows the design and analysis of a typical DSSS baseband transmitter. All electrical data and the transmitter layout are included in this chapter.

9.2 Brief Description of Transistor Design and Modeling

Both NMOS and PMOS transistors are enhancement type with uniform, undoped silicon bodies and metal gates constructed perpendicular to the SOI substrate as discussed earlier in Chapter 1. Device simulations are performed with Silvaco's 3-dimensional ATLAS device simulator with a 1 V power supply voltage.

The optimal SNT body dimensions of 2 nm radius and 10 nm channel length are determined according to the minimum intrinsic transient times and minimum static and dynamic power dissipations for each NMOS and PMOS transistor discussed in Chapter 3. The BSIMSOI device models also generated in Chapter 3 are used for all the circuit simulations in this chapter.

A. Bindal, S. Hamedi-Hagh, *Silicon Nanowire Transistors*,
DOI 10.1007/978-3-319-27177-4_9

9.3 DSSS Baseband Transmitter

The Direct Sequence Spread Spectrum (DSSS) transmitter is implemented according to the schematic in Fig. 9.1. In this figure, the baseband portion of the transmitter is composed of four blocks: an 8-Phase Shift Keying (8-PSK) modulator, a PN generator, a binary mapper and two bit multipliers.

9.3.1 Overall Operation of the Transmitter

Figure 9.2 illustrates how the symbol (010) is processed through DSSS baseband transmitter using CommSim communications simulator. The modulator and PN generator clocks are also included in the simulation results to show the data processing sequence that ends with an analog output.

When a valid 3-bit serial input, (010), is fed to the input of the 8-PSK modulator, the modulator generates +0.707 In-phase (I) and −0.707 Quadrature (Q) components of the symbol (010) according to the unitary constellation map in Fig. 9.1. Once generated, the component values are held constant for three modulator clock periods while the next symbol is fed to the modulator. The PN generator produces a repeating 15-bit (000100110101111) chip sequence at the Cpn node with a frequency five times faster than the modulator clock frequency and each chip in the sequence is multiplied by the I and Q components of the symbol to produce the frequency spread of MI and MQ values as shown in Fig. 9.2.

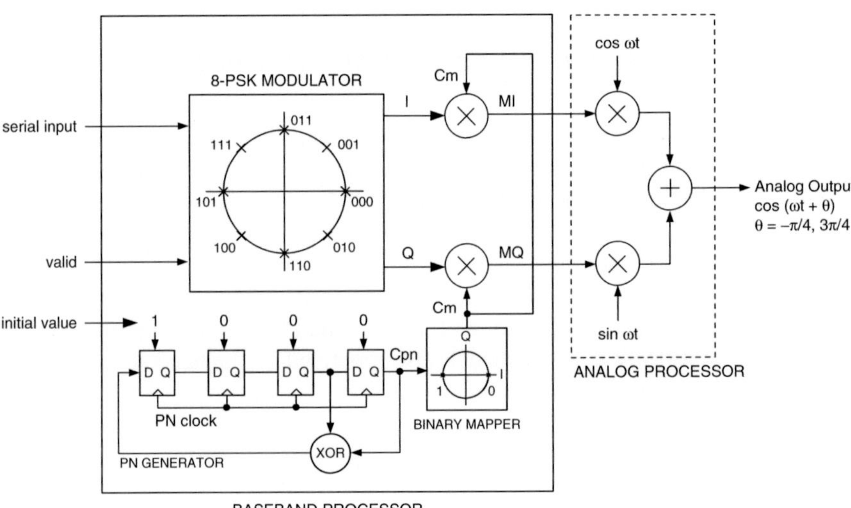

Fig. 9.1 DSSS baseband transmitter top level architecture

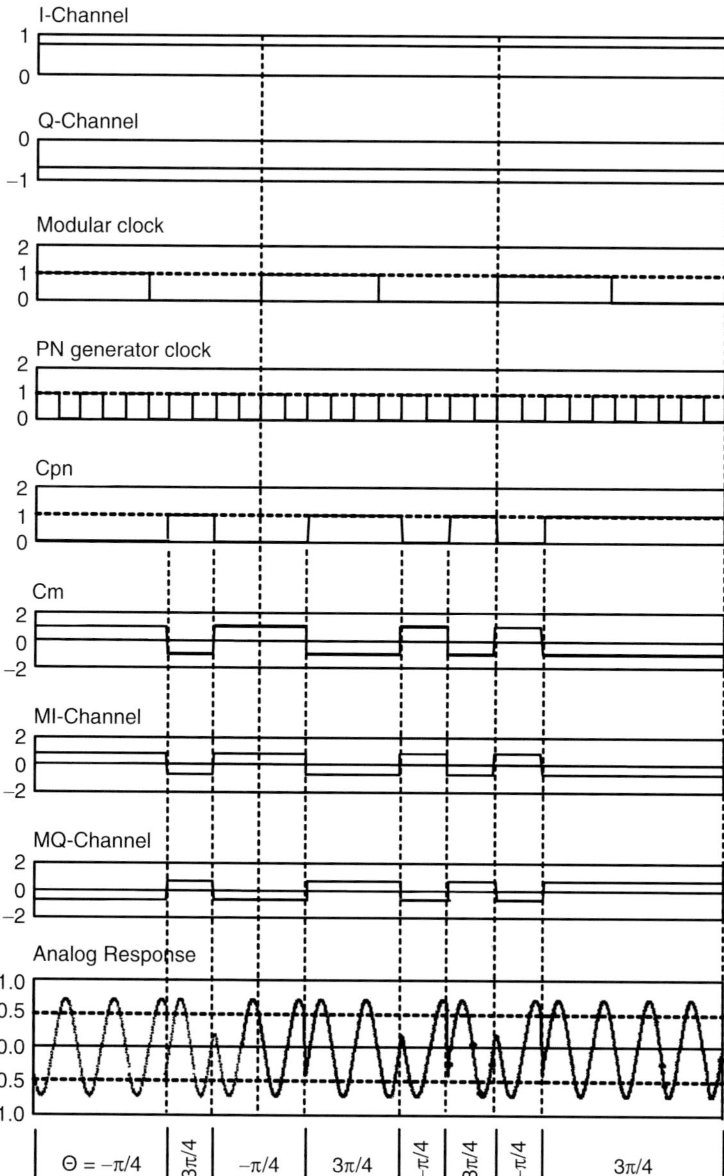

Fig. 9.2 CommSim symbol processing in the DSSS transmitter

The initial values of MI = +0.707 and MQ = −0.707 in this figure correspond to a phase shift of −π/4 according to the constellation circle. Therefore, a sine wave with a phase shift of −π/4 is generated during the first three chip periods. Then in the fourth chip, MI and MQ values change to −0.707 and +0.707, which causes the phase of the analog response to shift to 3π/4 and results in a discontinuity as shown

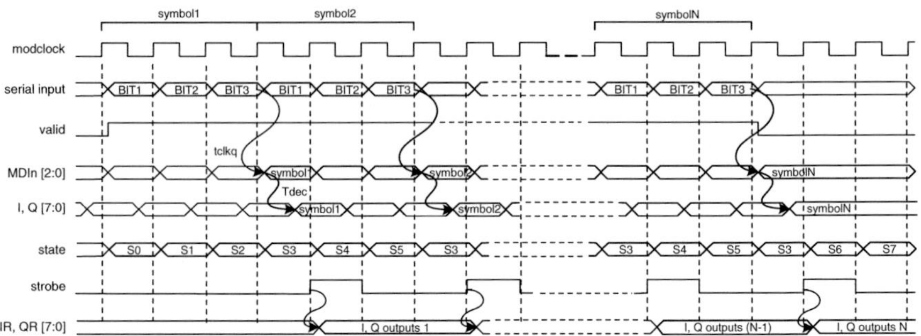

Fig. 9.3 Modulator timing diagram

in Fig. 9.2. The rest of the chip sequence reveals phase shifts between $-\pi/4$ and $3\pi/4$ and causes discontinuities in the analog output due to different MI and MQ values during each chip period.

9.3.2 8-PSK Modulator

The 8-PSK modulator generates I and Q components of a symbol as a function of serial input bit stream as shown in Fig. 9.3. The three valid input bits defining the first symbol is fed to the shift register of the modulator in Fig. 9.4. Within the forth clock period, the modulator decoder determines I and Q values of the symbol according to the constellation circle after a decoder delay, Tdec. In order to keep the symbol value constant for the next 3 clock cycles, a strobe clock is generated by the state machine in Fig. 9.5, which produces a high strobe signal during the state S4 as long as serial inputs are valid. When the last set of valid input bits is fed to the modulator, a final strobe signal is generated by the state S6 in order to latch and output the last symbol. While the registered I and Q symbol components, IR [7:0] and IQ [7:0], are held constant for three modulator clock periods for processing, the modulator continues to work in a pipeline fashion and simultaneously receives valid input bits to form the next symbol. In this design, I and Q components are 8-bit signed numbers; negative values are calculated using 2s complement addition.

9.3.3 PN Generator

The frequency-spreading element of the DSSS transmitter is the PN generator which subdivides a symbol into a 15-bit chip sequence according to its order [1]. In this system, a PN generator of the fourth order with four serial flip-flops and an XOR gate in Fig. 9.1 and also in Fig. 9.4 produces a repeating sequence of $2^4 - 1 = 15$ chips at Cpn. This sequence is dependent on the initial value loaded into

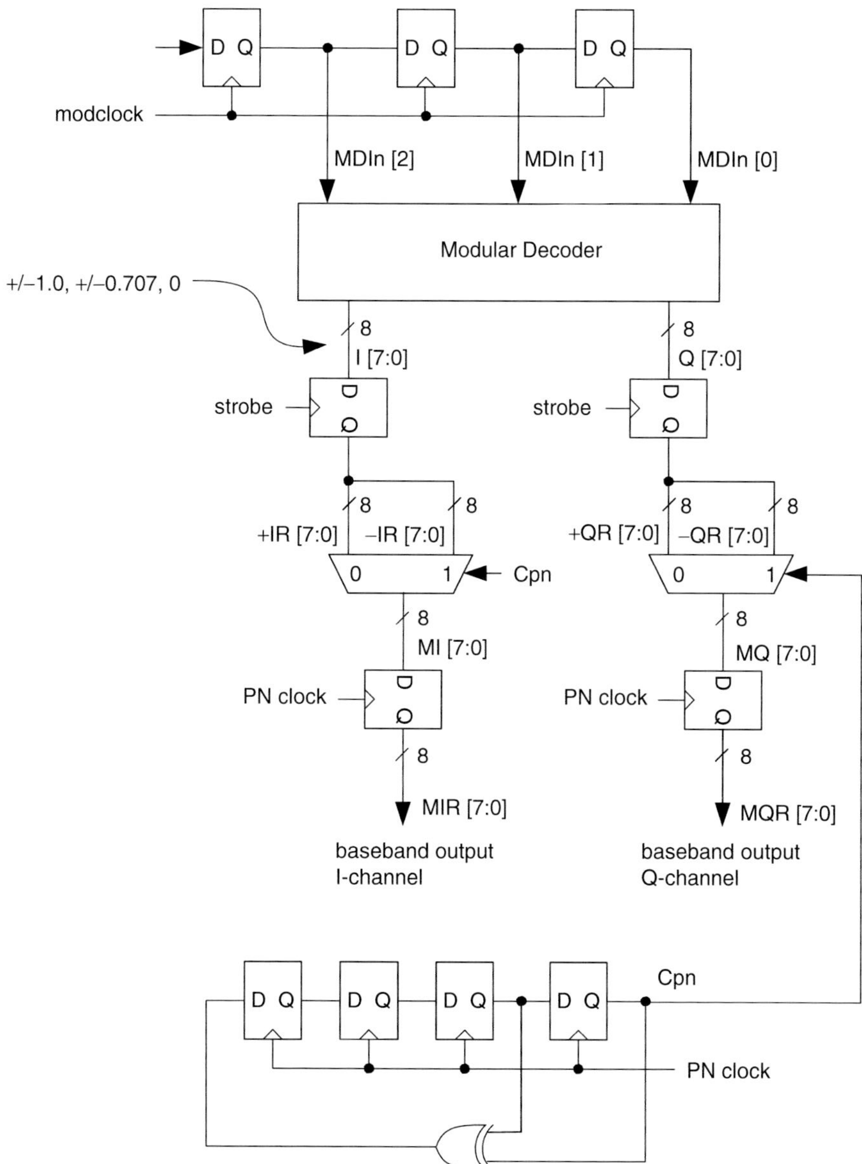

Fig. 9.4 DSSS transmitter functional block diagram

the PN generator where logic 1 is stored at the most significant flip-flop position and logic 0 is stored at the least significant flip-flop position of the PN generator.

The PN clock used in the PN generator is also used to generate the modulator clock as shown in Fig. 9.6. The PN clock frequency is selected to be five times as high as the modulator clock frequency to contain all 15 chips within three modulator clock periods.

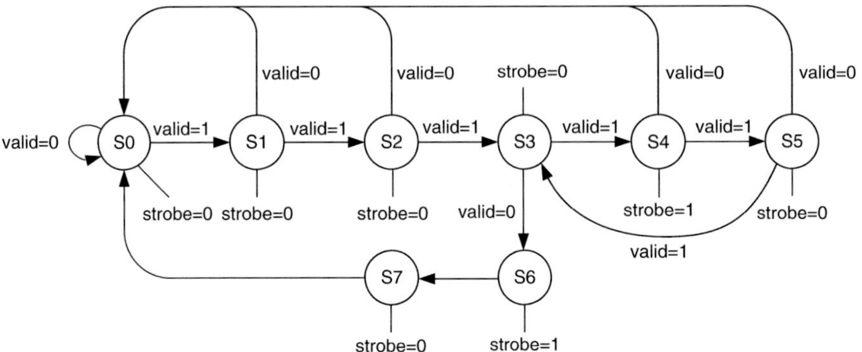

Fig. 9.5 The state machine generating the strobe clock from the modulator clock

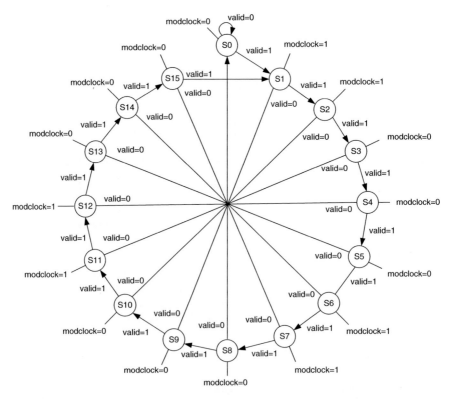

Fig. 9.6 The state machine generating the modulator clock from the PN clock

9.3.4 *Binary Mapper and Bit Multipliers*

The binary mapper in Fig. 9.1 produces +1 or −1 at the Cm node every time it receives logic 0 or logic 1 from the PN generator, respectively. I and Q channel values from the modulator are then multiplied by each chip at the Cpn node, and 8-bit wide, frequency-spread MI and MQ outputs are produced at MI [7:0] and MQ [7:0] nodes as shown in Fig. 9.4. Registered MI and MQ outputs, MIR [7:0] and MQR [7:0], are subsequently sent to the analog processor.

9.4 Circuit Simulations

9.4.1 *Clock Generation Circuits*

The Moore-type state machines in Figs. 9.6 and 9.7a are interconnected to show how each clock is generated from a single clock. The modulator clock generator is composed of FB-PN and FF-PN paths and operates with a high frequency clock, pnclockgen, to produce a secondary clock, modclockgen. Similarly, the strobe clock generator is composed of FB-mod and FF-mod paths and operates with modclockgen to produce strobe clock. The delay circuits between pnclockgen and PN clock, and between modclockgen and modclock, simply align all 3 clocks with respect to the original clock, pnclockgen, and eliminate the possibility of any hold time violation in the transmitter. Figure 9.7b shows the alignment of all system clocks within 1 ps.

9.4.2 *Maximum Critical Paths*

During the circuit analysis, four critical paths have emerged in the baseband transmitter. One of these paths, designated as path A around the feedback loop of the modulator clock generator (FB-PN) in Fig. 9.7a, dominates all the other critical paths B, C, and D shown in Figs. 9.7a, 9.8 and Table 9.1, and ultimately determines the PN clock frequency. Path A consists of two inverters, one 3-input and four 2-input NAND gates all connected in series as shown in Fig. 9.9a. It produces 10.2 ps pnclockgen-Q delay due to a large capacitive load at the PS [0] node and 17.3 ps total gate delay between the PS [0] and NS [3] nodes as shown in Fig. 9.9b. Since the setup time of the flip-flop is 3.5 ps, the total delay between two pnclockgen boundaries becomes $10.2 + 17.3 + 3.5 = 31$ ps as shown in Table 9.1. Adding 10 % RC parasitic delay overhead to this value brings up the PN clock period to be approximately 35 ps. Therefore, the modulator clock period containing

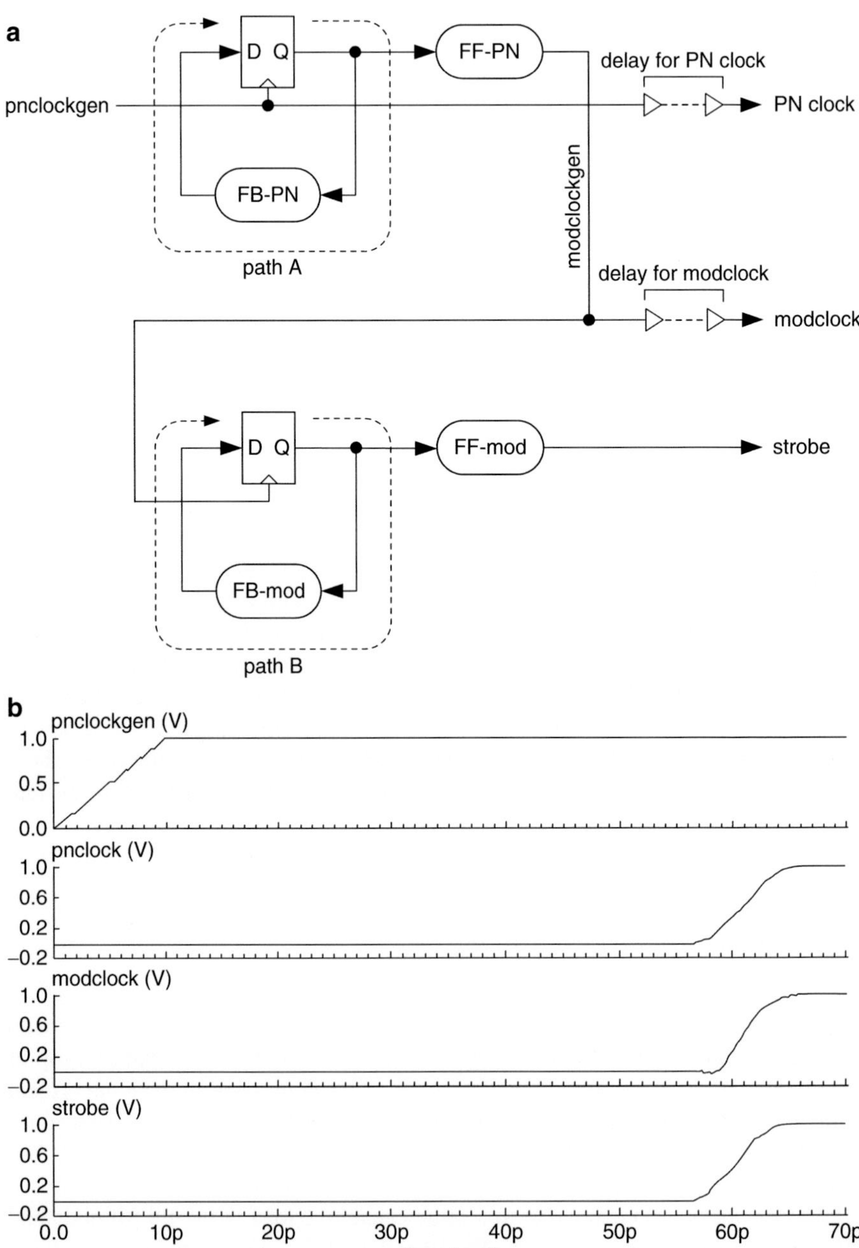

Fig. 9.7 (**a**) Critical paths in the clock generation circuits. (**b**) Generation of PN clock, modclock, and strobe

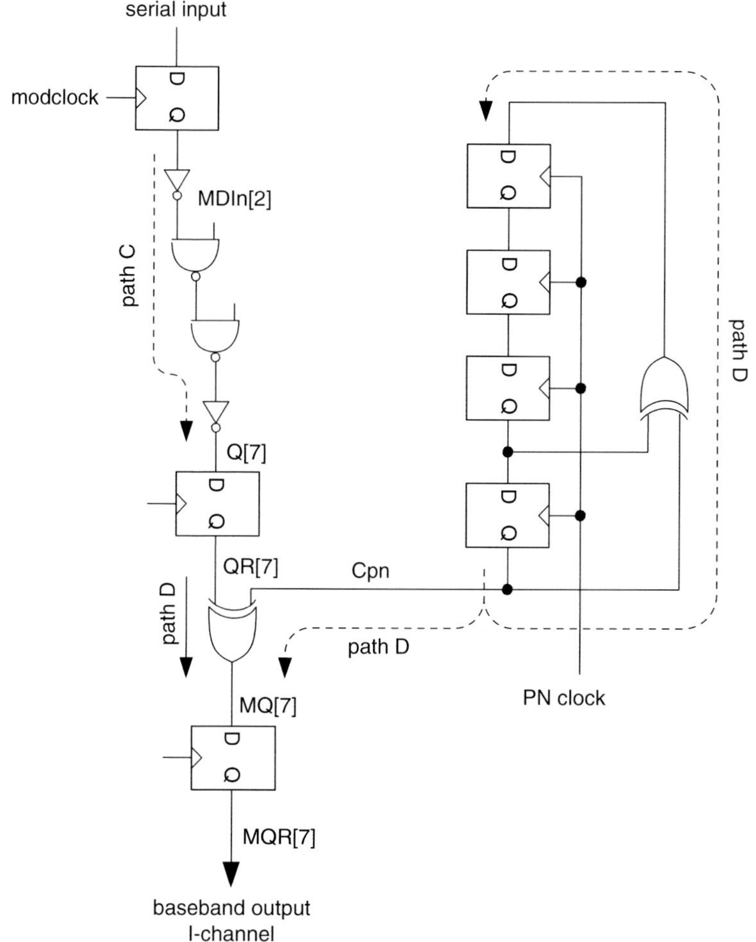

Fig. 9.8 Critical data paths in the modulator, bit mapper, and PN generator

Table 9.1 Critical path properties in DSSS transmitter

Critical	Clk-Q delay	Logic delay (ps)	Setup time (ps)	Total delay (ps)
Path A	10.2	17.3	3.5	31.0
Path B	9.1	14.2	3.5	26.8
Path C	5.8	7.9	3.5	17.2
Path D	5.8	2.2	3.5	11.5

5 chips becomes 175 ps and the strobe period containing all 15 chips or a single symbol becomes 525 ps. This translates approximately 28.6 GChips/s chip rate at the output of the baseband transmitter and makes this design suitable for any high-speed wireless application.

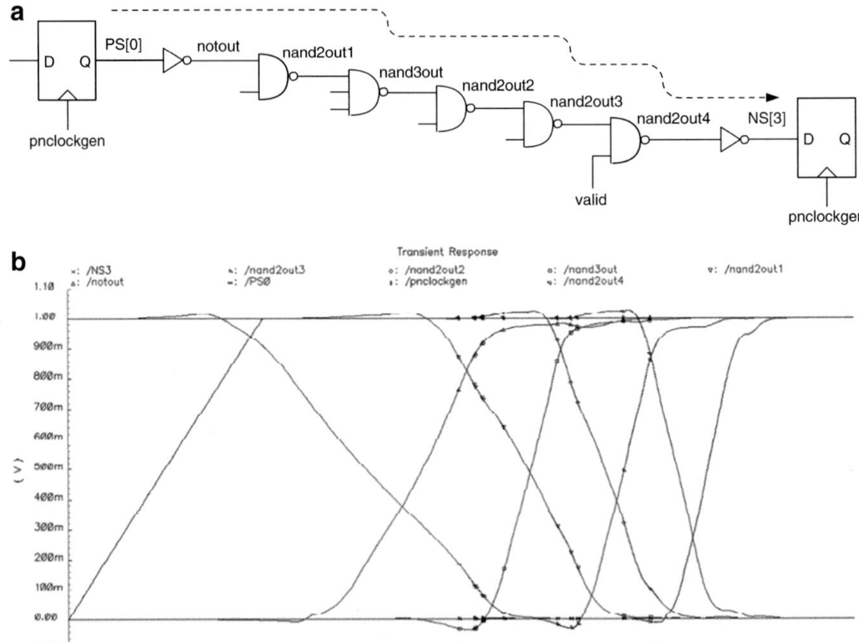

Fig. 9.9 (**a**) Logic string defining the critical path though FB-PN. (**b**) Circuit simulation results at various nodes of the feedback path, FB-PN

9.4.3 Minimum Critical Paths

Minimum critical paths are important in order to prevent hold violations in case of a clock shift. Both the PN generator and the shift register in the modulator have zero-delay paths between two flip-flop boundaries. Since the minimum clock-Q delay of a flip-flop is 4.6 ps for a fan-out of one and its hold time is 0.8 ps, the PN clock can be permitted to shift less than 3.8 ps to avoid hold time violation.

9.4.4 Parasitic RC Extraction

Two critical implementation issues of this design are to determine the layout design rules and the dielectric thickness between two metal layers to reduce the effects of parasitic resistance and capacitance on gate propagation delays. NMOS and PMOS transistors are deliberately built on circular silicon islands of SOI substrate as discussed in Chapter 1 to prevent latch-up and eliminate N-well and P-well junction capacitances.

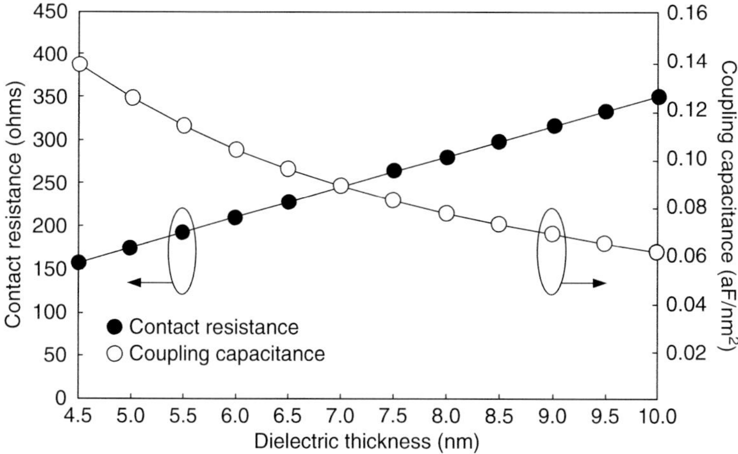

Fig. 9.10 Contact resistance of via openings and inter-metal layer coupling capacitance

Metal wire width for interconnects is kept at 4 nm for all metal layers. Since earlier experimental studies [2, 3] suggest that 10 nm metal thickness is necessary to avoid grain boundary formation and maintain a continuous metallic film, this thickness value is used for transistor gates as well as all metal interconnects. Wire interconnect spacing is assumed to be 4 nm to limit the value of coupling capacitance to 8.6×10^{-2} aF/nm^2 between two adjacent wires (height = 10 nm, width = 4 nm). The variations in contact resistance and coupling capacitance as a function of dielectric thickness are shown in Fig. 9.10.

Determining the value of the dielectric thickness between two metal layers is a trade-off between the coupling capacitance of overlapping interconnects and the contact resistance resulted from a 2×2 nm contact opening. As the dielectric thickness increases, the coupling capacitance between two metal interconnects reduces at the expense of increasing contact resistance. In Fig. 9.10, a compromise is made to select the dielectric layer thickness to be 7 nm such that minimal values of coupling capacitance (0.09 aF/nm^2) and a contact resistance (250 Ω) can be obtained. However, this dielectric thickness resulted in less than 3 fs parasitic delay on maximum critical paths; therefore, its effect is ignored from circuit simulations.

Inter-metal and intra-metal coupling capacitances are computed using Ansoft's capacitance calculator which reveals coupling, fringing, and area components of a single interconnect with respect to neighboring wires. The interconnect wire resistivity is determined by extrapolating the results of an earlier theoretical study [4] for the wire dimensions used in this study.

9.4.5 Power Consumption

The average power consumption of the DSSS baseband transmitter is found to be 198.5 μW at 5.7 GHz modulator clock frequency. Modulator and strobe clock generation circuits consume 93.8 μW while the remaining transmitter datapath including the modulator decoder, the PN generator, the bit mapper, and the two multipliers consumes an average of 104.7 μW. The total transmitter power dissipation changes between 188.1 μW for the symbol (111) and 213.4 μW for the symbol (010) according to the symbol being processed in Fig. 9.11. Each power consumption value in Fig. 9.11 is determined by averaging the time-integrated dynamic current drainage during a symbol period.

While processing symbols, high current spikes have occurred at the positive edge of each modulator clock. One such example is shown in Fig. 9.12 for the highest power dissipating symbol (010). Here, high dynamic power consumption levels tapered off rapidly as the rest of the chips in the modulator clock period are processed. Therefore, one can conclude that high current drainage results from the state change in the modulator shift register and the subsequent switching activity in the modulator decoder. Clock generator circuits are independent of processed symbol and produce constant power consumption.

Figure 9.13 shows chip power consumption for the symbol (010). Power consumption values in this figure are time averages and calculated by integrating dynamic power supply current as a function of time and dividing the result by the chip period. The power dissipation spikes in Fig. 9.12 also manifest themselves in Fig. 9.13 by reaching 300 μW when chips 1, 6, and 11 are processed and the modulator decoder experiences full activity. The rest of the chips consume relatively lower power below 100 μW due to the local activity in the bit multipliers and binary mapper. Figure 9.13 also shows the power consumption in the clock generator circuits as a function of chip number. The power dissipation changes between 43.9 and 133.7 μW depending on the active feedback and feed-forward paths in the modclock and strobe state machines.

Table 9.2 compares the chip rate, power dissipation, and the technology figures of this study with recent baseband transceivers designs [5–8]. Silicon nanowire transistors

Fig. 9.11 Power consumption in the DSSS baseband transmitter for different symbols

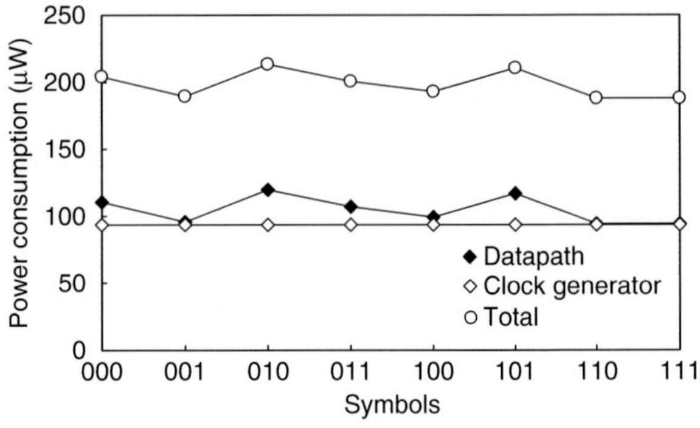

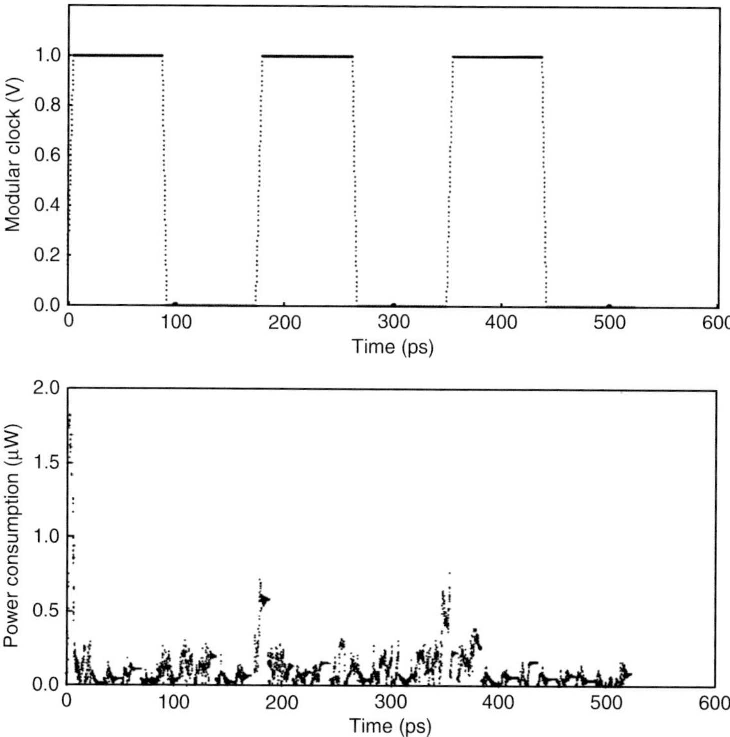

Fig. 9.12 Dynamic power consumption while processing symbol (010)

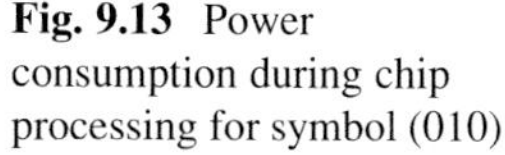

Fig. 9.13 Power consumption during chip processing for symbol (010)

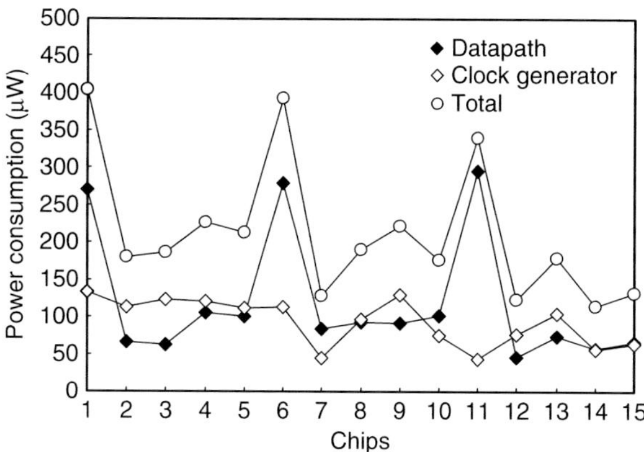

with 10 nm channel length and 2 nm radius deliver almost 30 times faster chip rate compared to the study in [8] even though the device properties of nanowire transistors were optimized for minimal power consumption. It is also interesting to note that Koyama et al. [8] report almost 50 times faster chip rate compared to the study in [7] even though the transceiver power dissipations in each study stay the same.

Table 9.2 Comparative study of this work with earlier designs

References	Chip rate	Power consumption	Technology
[5]	22 MChips/sec	390 mW[a]/850 mW[a]	0.8 μm/5 V supply
[6]	22 MChips/sec	1.2 W[b]	0.6 μm/3.3 V supply
[7]	22 MChips/sec	70 mW[a]/184 mW[a]	0.18 μm/1.8 V supply
[8]	1 GChips/sec	181 mW[b]	0.13 μm/1.2 V supply
This study	28.6 GChips/sec	198.5 mW[a]	10 nm/1 V supply

[a]Power dissipation in the transmitter
[b]Power dissipation in the transceiver

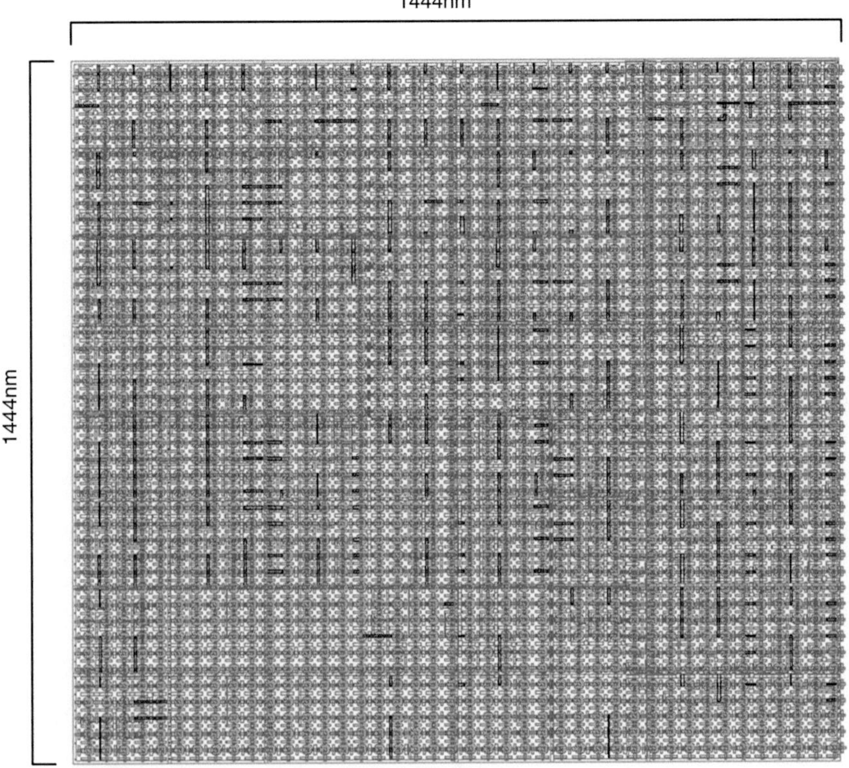

Fig. 9.14 DSSS baseband transmitter layout

9.4.6 Layout

Layout methodology used in this design is fabric-like, in the form of a crossbar configuration, as discussed in earlier chapters. In a crossbar structure, NMOS and PMOS transistors are placed in an alternate pattern that forces every NMOS transistor to have 4 neighboring PMOS transistors and vice versa.

With this fabric matrix configuration, a place-route CAD tool can be used to convert the entire circuit schematic into layout without any human interface. No custom design or ASIC approach is needed to create or characterize primitive or mega cells. Continuous lines are achieved by connecting pre-laid out metal interconnects with short metal strips. Figure 9.14 illustrates the result of fabric layout methodology for implementing the DSSS baseband transmitter. Over 90 % transistor utilization factor was achieved out of 1600 transistors used in this layout. The layout dimensions are 1444 nm by 1444 nm.

9.5 Summary

This chapter studies the possibility of using the SNT technology in a simple DSSS baseband transmitter and examines the results of the transmitter performance and power consumption. The transmitter contains four functional units: an 8-PSK modulator, a fourth order PN generator, a binary bit mapper and two bit multipliers. It generates In-phase and Quadrature outputs. The circuit simulations use the BSIMSOI nanowire transistor models developed in Chapter 3 and reveal a 31 ps worst-case delay between the two flip-flop boundaries with 28.6 GHz PN clock. Parasitic RC extractions add a maximum 3 fs delay on the designated critical paths, causing no setup time violations, and therefore ignored in the generation of the chip and modulator clocks. The resultant chip rate of 28.6 Gbits/s and the average power of 198.5 μW at a 1.9 GHz symbol frequency make this transmitter suitable for any handheld application. The layout was organized in a fabric-like fashion where NMOS and PMOS transistors are placed alternatively on a SOI substrate. The entire baseband transmitter layout occupies approximately 2.1 μm^2 of chip area.

References

1. Lee JS, Miller LE. CDMA systems engineering handbook. Artech House Publishers, ISBN: 0-89006-990-5
2. Kawamura M, Mashima T, Abe Y, Sasaki K (2000) Formation of ultra-thin continuous Pt and Al films by RF sputtering. Thin Solid Films 377–378:537–542
3. Liu HD, Zhao YP, Ramanath G, Murarka SP, Wang GC (2001) Thickness dependent electrical resistivity of ultrathin (<40 nm) Cu films. Thin Solid Films 384:151–156
4. Srivastava N, Banerjee K (2004) A comparative scaling analysis of metallic and carbon nanotube interconnections for nanometer scale VLSI technologies. Proc 21st Conf Int VLSI Multilevel Interconn Conf: 393–398
5. Wu JS, Liou ML, Ma HP, Chiueh TD (1997) A 2.6-V, 44-MHz all-digital QPSK direct-sequence spread-spectrum transceiver IC. IEEE Trans Solid-State Circuits 32(10):1499–1510
6. Chang HM, Sunwoo MH (1998) Implementation of a DSSS modem ASIC chip for wireless LAN. IEEE Workshop Signal Proc Sys: 243–252

7. Lin YH, Chan KU, Chang CJ, Chen TM, Lin YY, Kang HC (2005) A single-chip direct-sequence spread-spectrum CMOS transceiver for high performance, low cost 2.4-GHz cordless applications. Asian Solid State Circuits Conf: 253–256
8. Koyama A, Iwami H, Mizoguchi Y, Tashiro S, Nishiyama F, Yamagata T, Hashimoto Y, Takada M, Watanabe K, Iwasaki J, Suzuki M (2006) A DSSS UWB digital PHY/MAC transceiver for wireless ad hoc mesh networks with distributed control. IEEE Solid State Circuits Conf: 992–1001

Index

A. Bindal, S. Hamedi-Hagh, *Silicon Nanowire Transistors*,
DOI 10.1007/978-3-319-27177-4